Daily Warm-Ups

GEOGRAPHY

Kate O'Halloran

Level I

The classroom teacher may reproduce materials in this book for classroom use only.
The reproduction of any part for an entire school or school system is strictly prohibited.
No part of this publication may be transmitted, stored, or recorded in any form
without written permission from the publisher.

1   2   3   4   5   6   7   8   9   10

ISBN 0-8251-4490-6

Copyright © 2003

Walch Publishing

P. O. Box 658 • Portland, Maine 04104-0658

www.walch.com

Printed in the United States of America

Table of Contents

Introduction   iv

Geography Tools and Terms 1–19

The United States 20–38

Canada 39–50

Latin America 51–65

Western Europe 66–86

Eastern Europe 87–98

Northern Eurasia 99–113

Southwest Asia 114–125

Africa 126–145

South and East Asia 146–169

Australia and the Pacific World 170–180

Answer Key 181

Daily Warm-Ups: Geography

The *Daily Warm-Ups* series is a wonderful way to turn extra classroom minutes into valuable learning time. The 180 quick activities—one for each day of the school year—review, practice, and teach geography facts and concepts. These daily activities may be used at the very beginning of class to get students into learning mode, near the end of class to make good educational use of that transitional time, in the middle of class to shift gears between lessons—or whenever else you have minutes that now go unused. In addition to providing students with fascinating geographic information, they are a natural path to other classroom activities involving critical thinking.

Daily Warm-Ups are easy-to-use reproducibles—simply photocopy the day's activity and distribute it. Or make a transparency of the activity and project it on the board. You may want to use the activities for extra-credit points or as a check on the geographic and critical-thinking skills that are built and acquired over time.

We used many different types of resources to compile these activities and made every effort to use the most current data available. Because of the changing nature of human populations, the figures in some of the *Warm-Ups* will become outdated. These activities should still prove useful, however, in giving students practice in reading and interpreting data and will likely reflect current trends. You might wish to add extension activities in which students track down the most current data for specific topics. Such research helps to reinforce to students that human geography is not static.

However you choose to use them, *Daily Warm-Ups* are a convenient and useful supplement to your regular lesson plans. Make every minute of your class time count!

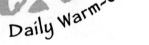

Geographer's Tools

It is impossible to show a three-dimensional shape on a flat surface. Because of this, all maps distort the way the world looks. Mapmakers use different ways to keep the distortion as small as possible. Each approach has advantages and disadvantages. The maps below show two common approaches to showing the earth on a flat map.

Mercator projection

Plate Carée

Compare the two types of maps. List as many advantages and disadvantages as you can for each type. Which one do you think shows the most accurate version of the world?

Water and Land

The terms below all refer to water and land coming together in some way. Write a short definition for each term.

alluvial plain

aquifer

bay

delta

fjord

floodplain

water table

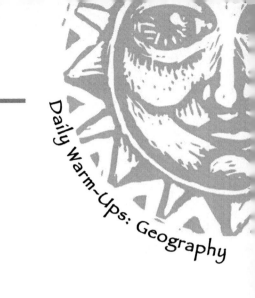

Maps

Making maps is a specialized skill called **cartography**. Each word below refers to cartography in some way. Write a short definition for each word.

altitude

cartogram

contour line

equator

latitude

longitude

map

meridian

parallel

projection

scale

topographic map

Which Map?

Maps are valuable tools. They help us show the three-dimensional world in a two-dimensional format. Like other tools, different types of maps are suited to different types of jobs.

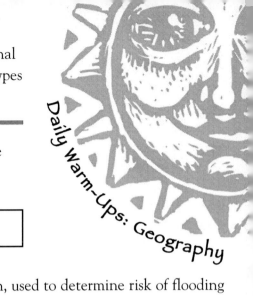

Several different types of maps are listed in the box. Choose the correct one for each project described below.

a. political map	b. physical map	c. thematic map

_____ 1. A study mapping watersheds in a region, used to determine risk of flooding

_____ 2. An article describing the countries in the area that was once Yugoslavia

_____ 3. A study comparing population densities in different states

_____ 4. A report about the different kinds of economic activity in a region

_____ 5. An itinerary for a traveler who plans to visit several different countries

_____ 6. A route plan for a hiker in the White Mountains of New Hampshire

_____ 7. A government study of natural resources

Weather vs. Climate

Weather and climate: they are part of our daily lives. Meteorologists study them both. But what is the difference between them?

It is sometimes said that "climate is what you expect, weather is what you get." Write a clear paragraph to explain what this statement means.

Fill in the Blanks: Map Terms

Choose the correct word from the box to complete each statement below.

equator	latitude
grid	longitude
hemisphere	

1. Lines of _____ are imaginary circles parallel to the equator.

2. Lines of _____ are imaginary semicircles that meet at the poles.

3. When it is summer in the northern _____, it is winter in the southern _____.

4. The _____ is at a latitude of 0°.

5. Lines of latitude and longitude form a _____ of imaginary lines that allows us to locate any place in the world.

© 2003 J. Weston Walch, Publisher

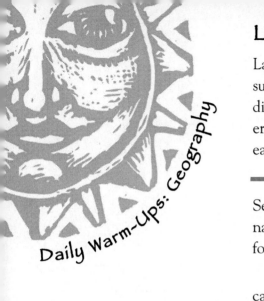

Landforms

Landforms are the physical features that make up the earth's surface. There are many kinds of landforms, created in many different ways—by the action of wind, water, and ice; through erosion; or through movement inside the earth, such as earthquakes and volcanoes.

Several different landforms are named below. For each landform, name at least one example. If possible, name the process that formed the physical feature you named.

canyon

delta

mountain

plain

plateau

valley

Migration

Every year, 60 million people migrate, or move from one place to another. Migration can be temporary or permanent. It can be within the same country or to another country. It can be voluntary, when the person chooses to migrate. Or it can be forced, when the person has no choice but to move.

People who study migration talk about push factors and pull factors. **Push factors** are factors that force a person to move. **Pull factors** are those that encourage a person to move. These factors usually fall into four general categories: economic, social, political, and environmental.

List as many push factors and pull factors for migration as you can. Try to think of factors in each of the four categories named above: economic, social, political, environmental.

World Biomes

The world can be divided into five large **biomes**, or places where life exists. Each large biome also includes several smaller subdivisions.

Name the five major biomes and as many subdivisions as you can.

Natural Disasters

Natural disasters can strike anywhere and can cause incredible damage. They can also take many forms. However, the hazards that cause most natural disasters can be grouped into two categories. **Atmospheric hazards** involve unusual or severe weather conditions. **Geologic hazards** are caused by processes within the earth itself.

Define and classify the natural disasters below (by writing **A** for atmospheric or **G** for geologic).

____ 1. avalanche

____ 2. blizzard

____ 3. drought

____ 4. earthquake

____ 5. flash flood

____ 6. hurricane

____ 7. landslide

____ 8. storm surge

____ 9. thunderstorm

____ 10. tornado

____ 11. tropical storm

____ 12. tsunami

____ 13. volcanic eruption

Pie Charts

Geographers use pie charts to show parts of a whole and how the parts relate to each other.

Look at the pie chart below. What does it show? How can you tell?

Once you are sure you know what the chart shows, write a descriptive label for the chart. Then write two or three clear sentences explaining what the chart shows.

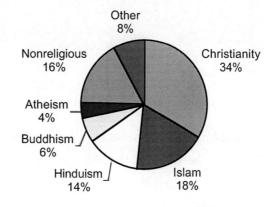

Population Density

Geographers use bar charts to compare a set of individual items. The bar's length shows how it compares to the other items in the set.

Look at the bar chart below. Then answer the questions that follow.

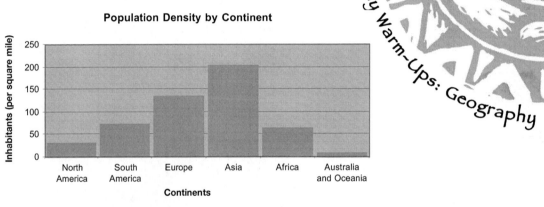

1. What do the bars of the chart show?
2. Is the information on the bars true for every square mile on each continent?
3. According to the chart, which continent is the most densely populated?
4. According to the chart, which continent is the least densely populated?
5. Do you think that this type of chart is a good way to show this information? Why or why not?

What's in a Name?

Many cities take their names from famous people. For example, Washington, D.C., was named for George Washington. But only a few countries are named after an individual.

Can you identify the country named after each person listed here?

1. Simón Bolívar
2. Christopher Columbus
3. Chief Nicarao
4. Johann von Liechtenstein
5. Abdul Aziz Al-Saud (Ibn Saud)
6. King Philip of Spain
7. Amerigo Vespucci

True or False?

Use your knowledge of geography to decide whether each of these statements is true or false. Write **T** for true or **F** for false on the line beside each statement.

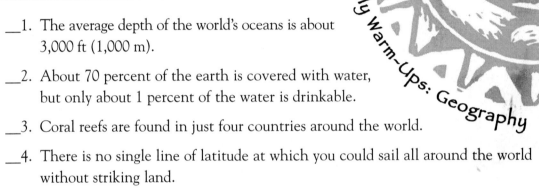

___1. The average depth of the world's oceans is about 3,000 ft (1,000 m).

___2. About 70 percent of the earth is covered with water, but only about 1 percent of the water is drinkable.

___3. Coral reefs are found in just four countries around the world.

___4. There is no single line of latitude at which you could sail all around the world without striking land.

___5. The names of all the continents end with the same letter that they start with.

___6. Because they are firmly based on solid rock, the continents do not move.

___7. The Ring of Fire is a circle of active volcanoes in the Pacific region.

___8. Glaciers look like solid mountains of ice, but they actually move, pushed forward by their enormous weight.

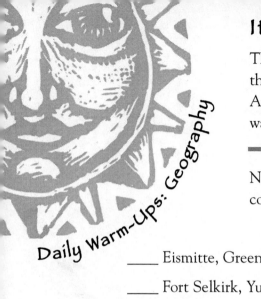

It's Cold out There

These are the 10 coldest places in the world. At the coldest of them all, a temperature of –128.6° F (–89.2° C) was recorded. At the least cold of the ten, a temperature of –70.0° F (–56.7° C) was recorded.

Number these places in order, from the coldest (1) to the least cold (10).

____ Eismitte, Greenland

____ Fort Selkirk, Yukon, Canada

____ Northice, Greenland

____ Oymyakon, Russia

____ Plateau Station, Antarctica

____ Prospect Creek, Alaska, U.S.A.

____ Rogers Pass, Montana, U.S.A.

____ Snag, Yukon, Canada

____ Verkhoyansk, Russia

____ Vostok, Antarctica

Some Like It Hot

These are the 10 hottest places in the world. At the hottest of them all, a temperature of 136°F (58°C) was recorded.

Number these places in order, from the hottest (1) to the least hot (10).

_____ Agha Jari, Iran

_____ Ahwaz, Iran

_____ Al'Aziziyah, Libya

_____ Araouane, Mali

_____ Death Valley, California, United States

_____ Ghudamis, Libya

_____ Kebili, Tunisia

_____ Tirat Tavi, Israel

_____ Tombouctou, Mali

_____ Wadi Halfa, Sudan

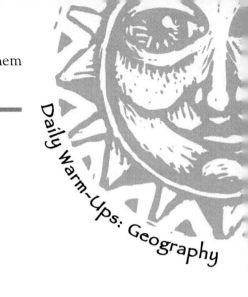

Daily Warm-Ups: Geography

16

People and the Environment

An important aspect of geography is the way people interact with the environment. This can take many forms. People may interact with the environment to get natural resources. Examples might be mining or logging. People may interact with the environment for pleasure, by hiking or swimming. Human–environment interaction may be as simple as raising an umbrella against the rain or as complex as planting trees to combat erosion.

Think about all the ways you interact with the environment. Consider all the ways people in your community interact with it. Then name as many specific ways as possible in which you and the rest of your community interact with your environment.

Languages of the World

Across the world, there are more than 2,700 languages. In addition, there are over 7,000 language **dialects**. These are regional variations of a language. The chart below shows the number of people who speak ten of the world's languages.

Look carefully at the chart. Then answer the questions.

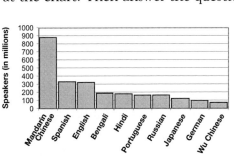

Source: Ethnologue: Languages of the World, 13th Edition, Barbara F. Grimes, Editor. © 1996, Dallas Summer Institute of Linguistics, Inc.

1. Which of these languages is spoken by the most people? About how many people speak it?
2. Which of these languages is spoken by the fewest people? About how many people speak it?
3. What is the relationship between the language on this chart with the most speakers and the language with the fewest speakers?

Geographic Extremes

Choose the correct word to complete each sentence below.

| Angel | Baikal | Nile | Russia |
| Australia | China | Pacific | Sahara |

1. The _____, at 4,145 mi (6,671 km), is the longest river in the world.

2. At 3,212 ft (979 m), _____ Falls, in Venezuela, is the world's highest waterfall.

3. The _____ Ocean is larger than any other ocean in the world.

4. Russia's Lake _____ is the world's deepest lake.

5. The _____ Desert in Africa, which covers more than 3 million sq mi (almost 8 million sq km), is the world's largest desert.

6. The world's largest island, _____, is also a continent.

7. With an area of 45,202,988 sq mi (17,075,200 sq km), _____ is the largest country in terms of area.

8. More than one billion people live in _____, making it the world's most populated nation.

© 2003 J. Weston Walch, Publisher

Geography Terms

Different regions have different characteristics. Because of this, each region has its own terminology—words that are used to describe the region.

Each word below describes an aspect of the geography of the United States. Write a clear definition explaining each word.

badlands	fault
cay	levee
chinook	mesa
coulee	piedmont
desert	

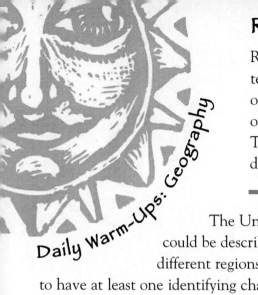

Regions

Regions are groups of places that have at least one common characteristic. One area may be part of many different regions, depending on which characteristic is being considered. Regions may be based on physical characteristics, such as climate, soil type, or landforms. They may be based on cultural characteristics, such as political divisions, economy, religion, or language.

Daily Warm-Ups: Geography

The United States includes many different regions, and many states could be described as belonging to several different regions. Name as many different regions of the United States as you can. Remember, each region needs to have at least one identifying characteristic, either physical or cultural. Which states are included in each region you named? Which of these regions does your state belong to?

Absolute Location

Absolute location is the exact position on Earth where a place can be found. No two places have the same absolute location. This is usually found by using imaginary lines—latitude and longitude—that mark positions on the surface of the earth.

Match each place named in the left column with its latitude and longitude in the right column.

Place

_____ 1. Anchorage, AK

_____ 2. Belfast, ME

_____ 3. Charleston, SC

_____ 4. Colorado Springs, CO

_____ 5. Oklahoma City, OK

Latitude and Longitude

a. 39° N, 105° W

b. 61° N, 150° W

c. 35° N, 98° W

d. 33° N, 80° W

e. 44° N, 69° W

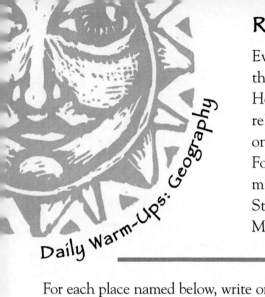

Daily Warm-Ups: Geography

Relative Location

Every place on Earth has just one specific location—the point on the latitude and longitude grid where that place can be found. However, a place can also have a **relative location**—location as it relates to other places. If we say that a place is north of, south of, on the shores of, and so on, then we are giving its relative location. For example, to describe the relative location of Philadelphia, we might say: "It is located in Pennsylvania, in the eastern United States, southwest of New York City, and northeast of Baltimore, Maryland."

For each place named below, write one or two sentences describing its relative location.

1. Boulder, Colorado
2. Chicago, Illinois
3. Miami, Florida
4. New Orleans, Louisiana
5. Seattle, Washington

© 2003 J. Weston Walch, Publisher

Where Am I?

Use the clues in this paragraph to identify the place being described.

Wild and remote, this state has been nicknamed "The Last Frontier." When U.S. Secretary of State William Seward bought the land from Russia in 1867, for two cents an acre, it was nicknamed "Seward's folly." Then gold was found in its rivers and streams, and it became an economic asset. It finally became a U.S. state in 1959.

This state is one of extremes. It is the largest state in the union, covering 656,425 sq mi (1,700,133 sq km). It has the coldest temperatures in the nation, and the lowest population density.

This state has the largest national forest in the United States, the Tongass National Forest. It also has the country's highest mountain, Mt. McKinley (Denali), which rises 20,320 ft (6,195 m) above sea level. In fact, 17 of the 20 highest peaks in the United States are found here.

The state's most important source of revenue comes from oil; it produces about one quarter of all oil produced in the United States. Other important industries are fisheries, wood and wood products, furs, and tourism.

Where am I?

Time Zones

The chart below shows the percentage of the U.S population that lives in each of five time zones.

Look at the chart carefully. Then answer the questions that follow.

Percent of U.S. Population in Each Time Zone

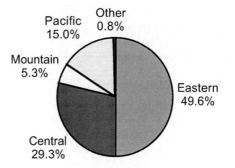

Pacific 15.0%
Other 0.8%
Mountain 5.3%
Eastern 49.6%
Central 29.3%

1. According to this chart, more people live in one U.S. time zone than in any other. Which one is that? What percentage of people live there?

2. The chart shows a percentage for "other." What areas might fall under this heading? Why? Write one or two clear sentences explaining your answer.

© 2003 J. Weston Walch, Publisher

Geography of License Plates

Each state has its own distinctive license plate. In fact, many states offer drivers a choice of plates. These plates are designed to show something about the state.

Some states do this in a visual way. For example, Colorado's license plate shows mountains silhouetted against the sky, for the Rocky Mountains. Other states print slogans on their license plates. These slogans are often based on the state's geography.

Daily Warm-Ups: Geography

Match each numbered state name with the letter of the correct license plate slogan.

____ 1. Alaska   a. Big Sky
____ 2. Arizona   b. Evergreen State
____ 3. Indiana   c. Great Lakes
____ 4. Michigan   d. The Last Frontier
____ 5. Minnesota   e. America's Dairyland
____ 6. Montana   f. The Crossroads of America
____ 7. Rhode Island   g. Green Mountain State
____ 8. Vermont   h. 10,000 Lakes
____ 9. Washington   i. Grand Canyon State
____ 10. Wisconsin   j. Ocean State

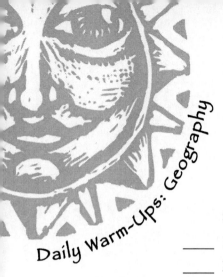

Strait and Narrow

A strait or channel is a narrow body of water that connects two larger bodies of water. There are approximately 200 straits or channels around the world. Some of them are in U.S. waters.

Match each strait in the left column with the bodies of water it connects, listed in the right column.

Strait

____ 1. Bering Strait
____ 2. Straits of Florida
____ 3. Strait of Juan de Fuca
____ 4. Hecate Strait
____ 5. San Pedro Channel

Bodies of Water

a. Pacific Ocean, Gulf of Santa Catalina
b. Gulf of Alaska, Pacific Ocean
c. Gulf of Mexico, Atlantic Ocean
d. Arctic Ocean, Bering Sea
e. Pacific Ocean, Puget Sound

What's in a Name?

A **toponym** is a place name. Place names come from a variety of sources. Some places are named after people. Some are named for a distinctive physical feature. Some tell something about the culture of the place.

Each place name listed below means something. Match each name with its meaning. Use your knowledge of geography to help you find the answers.

____ 1. Alaska   a. father of waters
____ 2. Connecticut   b. snowcapped
____ 3. Florida   c. feast of flowers
____ 4. Kentucky   d. great river
____ 5. Massachusetts e. great land
____ 6. Minnesota   f. at or about the great hill
____ 7. Mississippi   g. mountainous
____ 8. Montana   h. on a long river
____ 9. Nevada   i. sky-tinted water
____10. Ohio   j. land of tomorrow

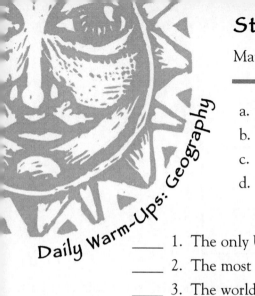

States of Extremes

Match each extreme described below with the state where it is found.

a. Hawaii   e. Virginia   i. Oregon
b. Alaska   f. Pennsylvania   j. New Hampshire
c. New Mexico   g. Michigan
d. California   h. Minnesota

____ 1. The only U.S. state with no natural lakes
____ 2. The most active volcanoes in the United States
____ 3. The world's shortest river, just 121 feet long
____ 4. The wettest weather in the United States
____ 5. The driest weather in the United States
____ 6. The northernmost point in the contiguous (48) states
____ 7. The world's highest ground-level wind speed
____ 8. The world's first oil well
____ 9. The highest capital city in the United States
____10. The state with the longest coastline in the U.S.A.

© 2003 J. Weston Walch, Publisher

Professional Sports Teams

Many professional sports teams have names that reflect the geography and culture of their home cities. The statements below describe the origins of eight professional team names. Choose the correct team name from the box for each statement.

Daily Warm-Ups: Geography

Boston Celtics	Minnesota Vikings	Seattle Mariners
Denver Broncos	New York Knicks	Texas Rangers
Detroit Pistons	San Francisco 49ers	

1. Many early settlers of this region came from Scandinavia.
2. In 1849 people flocked to this area, looking for gold.
3. This northwestern city is on an inlet of the Pacific Ocean.
4. This Rocky Mountain city was a center for cowboys and their horses.
5. During the 1800s, thousands of Irish immigrants came to this city.
6. This city is a center of the U.S. car industry.
7. A law enforcement unit of this name was started in 1823.
8. Dutch settlers who founded the first colony here wore knickerbockers.

© 2003 J. Weston Walch, Publisher

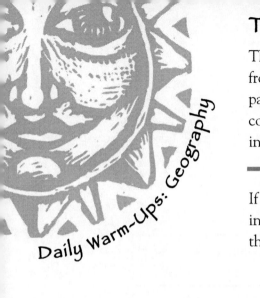

The Appalachian Trail

The Appalachian Mountains extend about 1,600 mi (2,574 km) from Canada south to Alabama. The Appalachian Trail, a hiking path that runs along the mountains, is the world's longest continuous mountain trail. Its endpoints are at Mount Katahdin in Maine and Springer Mountain in Georgia.

If you hiked the Appalachian Trail, starting in Georgia and ending in Maine, which states would you go through? Name as many of them as you can. (Hint: The trail passes through 14 states.)

Population Density

The average population density for the whole United States is 67 people per square mile (26 per square kilometer).

According to this table, in 1993, which state had almost the same population density as the entire United States?

	1993 Population	Square Miles of Land Area	Population Density (per sq mi)
Florida	12,679,000	58,664	233.0
Texas	18,031,000	266,807	67.5
Virginia	6,491,000	40,767	159.0

Population of Washington State

Geographers use line graphs to show how something changes over time. The slope of the line tells viewers at a glance whether the variable is getting bigger or smaller.

The line graph below shows the population of Washington State from 1890 to 2000. Look carefully at the graph. Then answer the questions that follow.

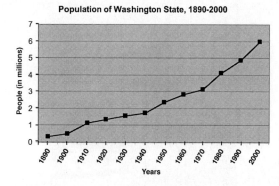

1. What was the approximate population of Washington in 1890?

2. What was the approximate population in 2000?

3. Which period on the graph shows the greatest proportional growth—that is, the greatest growth in relation to the earlier figure?

4. Which period on the graph shows the greatest actual growth—that is, the greatest jump between one date and the next?

Census 2000

Geographers use pie charts to show parts of a whole, and how the parts relate to each other. This pie chart is based on information from the 2000 U.S. Census for the state of Nebraska. What is the information being shown? How can you tell?

Once you are sure you know what the chart shows, write a descriptive label for the chart. Then write two or three clear sentences explaining what the chart shows.

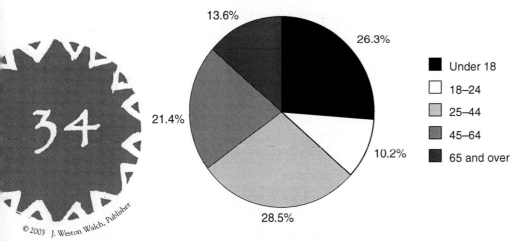

NAFTA

In 1994, the United States, Mexico, and Canada signed a trade agreement. The North American Free Trade Agreement, or NAFTA, created a huge free-trade area. Its goals included reducing barriers to trade among the three countries.

Look carefully at the graph. Then answer the questions that follow.

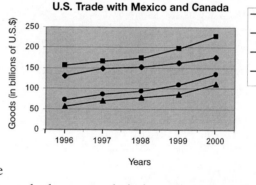

1. What does the graph show?

2. Trade balance between two countries means that imports and exports between them have about the same value. Does the graph show a trade balance between the United States and Canada? Between the United States and Mexico?

3. One of the goals of NAFTA was to increase trade between member countries. Based on the graph, do you think this goal is being met?

Multiple Choice

Choose the best answer for each question below.

1. A Pacific current has a warming effect on the southwestern coastal region of which high-latitude state?

    a. Alaska   b. Maine   c. California   d. New Hampshire

2. What kind of physical feature forms the southern boundaries of the states of Illinois, Indiana, and Ohio?

    a. mountain range   b. lake   c. river   d. plateau

3. In what central Great Plains state do the North and South Platte Rivers join to form the Platte River?

    a. South Dakota   b. Nebraska   c. Iowa   d. North Dakota

4. The largest oil-producing region in the United States is located in the South. In what state is it found?

    a. West Virginia   b. North Carolina   c. Florida   d. Texas

5. The two U.S. states that are located farthest away from the Northeast Region are

    a. California and Texas   b. Alaska and Hawaii
    c. Washington and Oregon   d. Alaska and Washington

True or False?

Use your knowledge of the region to decide whether each of these statements is true or false. Write **T** for true or **F** for false on the line beside each statement.

____ 1. The driest place in the United States is Mount Waialeale on the Hawaiian island of Kauai.

____ 2. St. Augustine, Florida, is the oldest continuously inhabited city in the United States.

____ 3. No U.S. cities have the same name as any state.

____ 4. The northernmost point in the contiguous (lower 48) states is in Maine.

____ 5. A pipeline brings oil 800 mi (1,280 km) from oil fields in northern Alaska to the port of Valdez.

____ 6. A state park in Georgia consists of half a million acres of swamp (2,000 sq km), the Okefenokee Swamp Park.

____ 7. The letters Q, X, and Z do not appear in the name of any U.S. state.

____ 8. The northernmost U.S. state capital is Juneau, Alaska.

____ 9. The only active diamond mine in the United States is in Arkansas.

____ 10. With an area of just 1,214 sq mi (1,953.7 sq km), Rhode Island is not just the smallest state; it is also smaller than any country in Europe.

Fill in the Blanks

Choose the correct word to complete each sentence below.

| Colorado | Kentucky | Minnesota | Nevada |
| Connecticut | Mexico | Missouri | Pennsylvania |

1. The Mississippi River rises in Lake Itasca, _____.

2. Between Florida and Texas, the United States is bordered by the Gulf of _____.

3. Independence, _____, was the eastern terminus of the Santa Fe and Oregon trails.

4. The driest state in the nation, _____ has an average annual rainfall of only about seven inches.

5. Founded by William Penn, _____ was once a center of the steel industry.

6. The first region west of the Allegheny Mountains to be settled by American pioneers, _____ is still known today for fine horses.

7. _____, nicknamed The Nutmeg State, is the insurance capital of the United States.

8. Pike's Peak is just one of more than 1,000 peaks in _____ that are over 10,000 ft (3,000 m) high.

© 2003 J. Weston Walch, Publisher

Regions

Canada is a land of broad expanses. Stretching from the Atlantic to the Pacific, it covers an area of 3,851,809 sq mi (9,976,140 sq km).

Canada is divided into ten provinces and three territories. Name them all.

Absolute Location

Absolute location is the exact point on Earth where a place can be found. No two places have the same absolute location. This is usually found by using imaginary lines—latitude and longitude—that mark positions on the surface of the earth.

Match each place named in the left column with its latitude and longitude in the right column.

Place	Latitude and Longitude
___ 1. Calgary, Alberta	a. 45° N, 66° W
___ 2. Ottawa, Ontario	b. 49° N, 123° W
___ 3. Québec City, Québec	c. 51° N, 114° W
___ 4. St. John, New Brunswick	d. 45° N, 76° W
___ 5. Vancouver, British Columbia	e. 47° N, 71° W

Relative Location

Every place on Earth has just one absolute location—the point on the latitude and longitude grid where it can be found. However, a place can also have a **relative location**. This means where it is relative to other places. If we say a place is north of, south of, or on the shores of, we are giving its relative location. For example, we might describe the relative location of Ottawa like this: "It is in Ontario, in eastern Canada, west of Montréal, and northeast of Toronto."

For each place named below, write one or two sentences describing its relative location.

1. Edmonton, Alberta
2. Iqaluit, Nunavut
3. Montréal, Québec
4. Toronto, Ontario
5. Winnipeg, Manitoba

Where Am I?

Use the clues in this paragraph to identify the place being described.

The Pacific Ocean bounds it on the west, the Rocky Mountains on the east. It is dotted with lakes, and scattered islands cling to its coast. Some of these islands are nestled in fjords cut into the Coastal Mountains. East of the Coastal Mountains lies a rolling upland of forests, grasslands, and lakes. Farther east again lie the heights of the Rocky Mountains, with peaks more than 13,000 ft (4,000 m) high. Just over 12 percent of Canada's population lives here, largely in the cities of Victoria and Vancouver.

Most of the province's economy is based on its natural resources. Vast forests cover more than half the land; their trees are used for lumber, paper, and other wood products. Mining is another important part of the economy. Copper, gold, and zinc are all found here. The province has valuable reserves of coal, petroleum, and natural gas. Agriculture, fishing, and tourism also add to the economy.

Where am I?

One-of-a-Kind

Match each feature described below with the place where it is found.

a. Bay of Fundy, Nova Scotia
b. Hudson Bay
c. Manitoulin Island, Ontario
d. Montréal, Québec
e. Ontario
f. Ottawa, Ontario
g. Queen Elizabeth Islands, Nunavut
h. Toronto, Ontario
i. Vancouver, British Columbia
j. West Edmonton, Alberta

Daily Warm-Ups: Geography

_____ 1. The CN Tower, the world's tallest self-supporting structure
_____ 2. The world's greatest tides, with a rise and fall as great as 40 ft
_____ 3. Largest French-speaking city in the Western Hemisphere
_____ 4. Birthplace of the only European royal princess born in North America
_____ 5. World's largest shopping mall
_____ 6. North Magnetic Pole—where the needle of a compass points
_____ 7. The only province that borders all four Great Lakes that touch Canada
_____ 8. World's largest bay, as measured by shoreline length
_____ 9. World's largest island on a lake
_____ 10. The world's narrowest building, just 4 ft, 11 in. wide and 96 ft long

© 2003 J. Weston Walch, Publisher

Population Change

The chart below shows the population of each of Canada's provinces in two years, 1996 and 2001. The last column shows the percentage that the population increased or decreased.

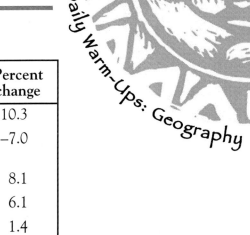

Study the chart. Then answer the questions that follow.

Province	2001	1996	Percent change
Alberta	2,974,807	2,696,826	10.3
Newfoundland and Labrador	512,930	551,792	−7.0
Nunavut	26,745	24,730	8.1
Ontario	11,410,046	10,753,573	6.1
Québec	7,237,479	7,138,795	1.4

1. Which province had the greatest population in 2001?
2. Which province had the smallest population in 2001?
3. Which province had the greatest increase in population between 1996 and 2001?
4. Does this table give you enough information to say which province has the greatest population density—that is, the most people per square mile?

Strait and Narrow

A strait or channel is a narrow body of water that connects two larger bodies of water. There are approximately 200 straits or channels around the world. Some of them are in Canadian waters.

Match each strait in the left column with the bodies of water it connects, listed in the right column.

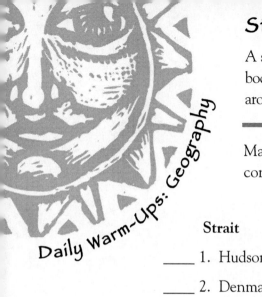

Strait

____ 1. Hudson Strait

____ 2. Denmark Strait

____ 3. Strait of Belle Isle

____ 4. Cabot Strait

____ 5. Davis Strait

Bodies of Water

a. Atlantic Ocean, Gulf of St. Lawrence

b. Greenland Sea, Atlantic Ocean

c. Hudson Bay, Labrador Sea

d. Baffin Bay, Labrador Sea

e. Atlantic Ocean, Gulf of St. Lawrence

Climate Severity

Environment Canada, a department of the Canadian government, has developed a climate severity index. This index is used to rate how severe a location's climate is. To do this, the index looks at a number of factors. These include

- extremes of heat or cold, wetness or dryness, windiness
- poor air quality
- continuous darkness or daylight
- long periods of rain, snow, fog, or restricted visibility
- severe weather such as thunderstorms, blizzards, blowing snow, windstorms, tornadoes, hailstorms, and freezing rain
- limitations on outdoor mobility

Once these factors have been considered, a location is given an index rating. A low rating means that the climate is very mild. (Victoria, BC, has a rating of 13.) A rating near 100 means that the climate is very severe. (Isachsen weather station, in the Northwest Territory, has a rating of 99.)

What does this kind of rating system tell you about Canada's climate? Write a clear paragraph to explain your answer.

Daily Warm-Ups: Geography

What's in a Name?

A **toponym** is a place name. Place names come from a variety of sources. Some places are named after people. Some are named for a distinctive physical feature. Some tell something about the culture of the place.

Each place name listed below means something. Match each name with its meaning. Use your knowledge of geography to help you find the answers.

____ 1. Canada   a. muddy water

____ 2. Ottawa   b. female grizzly bear

____ 3. Québec   c. narrow channel

____ 4. Toronto   d. big village

____ 5. Winnipeg   e. early berries

____ 6. Iqaluit   f. great river

____ 7. Labrador   g. landowner

____ 8. Saskatoon   h. to trade

____ 9. Kelowna   i. place of fish

____ 10. Yukon   j. where there are trees standing in the water

© 2003 J. Weston Walch, Publisher

Multiple Choice

Choose the best answer for each question below.

---

1. What is the largest island in Canada?

    a. Prince Edward Island   c. Baffin Island
    b. Banks Island   d. Newfoundland

2. On which three oceans does Canada have coastlines?

    a. Pacific, Atlantic, and Indian   c. Atlantic, Pacific, and Arctic
    b. Atlantic, Arctic, and Indian   d. Pacific, Indian, and Arctic

3. Four of the five Great Lakes are in both the United States and Canada. Which lake is not in Canada?

    a. Lake Michigan   b. Lake Superior   c. Lake Ontario   d. Lake Erie

4. What is the only walled city in North America north of Mexico?

    a. Montréal   b. Québec   c. Vancouver   d. Winnipeg

5. Which city is almost midway between the Atlantic and Pacific coasts?

    a. Ottawa   b. Toronto   c. Calgary   d. Winnipeg

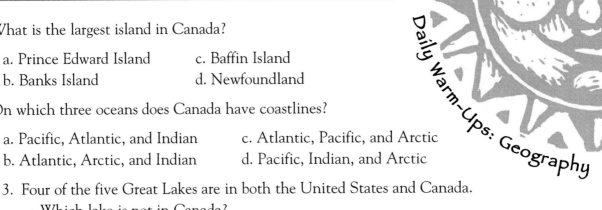

Daily Warm-Ups: Geography

48

True or False?

Use your knowledge of the region to decide whether each of these statements is true or false. Write **T** for true or **F** for false on the line beside each statement.

___ 1. Twenty-seven U.S. states have land north of Canada's southern-most point.

___ 2. The 4,724-mi (7,604-km) Trans-Canada Highway is the world's longest national highway.

___ 3. Canada has fewer lakes than any other country in the Northern Hemisphere.

___ 4. Canada's border with the United States is 5,525 mi (8,891 km) long.

___ 5. Since Toronto has no direct access to the Atlantic, all goods from the city must be shipped overland.

___ 6. Canada has the longest coastline of any country.

___ 7. Canada covers three time zones.

___ 8. The climate in British Columbia is the most severe in Canada.

___ 9. Canada has mountains named for Santa's reindeer—Dasher, Dancer, Prancer, Vixen, Comet, Cupid, Donner, and Blitzen.

___10. Few people in Canada live near the U.S. border.

© 2003 J. Weston Walch, Publisher

Fill in the Blanks

Choose the correct province or territory name to complete each sentence below. Some names will be used more than once.

Alberta	Nunavut
British Columbia	Ontario
Newfoundland and Labrador	Yukon
Nova Scotia	

1. Canada's southernmost point is at Middle Island, _____.
2. The province of _____ has the least farmland of any Canadian province.
3. Canada's highest mountain, Mount Logan, is in _____ Territory.
4. The highest waterfall in Canada is Della Falls, in _____.
5. Buffalo Wood National Park, in _____ and the Northwest Territories, is home to the largest herd of bison in the world and is the only nesting site of the endangered whooping crane.
6. The driest place in Canada is in Arctic Bay, _____.
7. With 14,423 mi (23,212 km) of coastline, the province of _____ has the longest coastline of any Canadian province.
8. The province of _____ consists of a peninsula and a fairly large island.

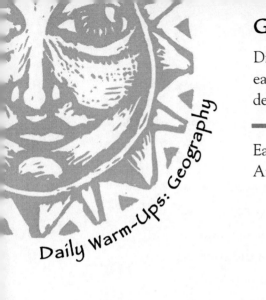

Geography Terms

Different regions have different characteristics. Because of this, each region has its own terminology—words that are used to describe the region.

Each word below describes an aspect of the geography of Latin America. Write a clear definition of each word.

altiplano	pampero
chaparral	páramos
cordillero	rain forest
El Niño	rain shadow
llano	sierra
pampas	

Regions

Regions are groups of places that have at least one common characteristic. One area may be part of many different regions, depending on which characteristic is being considered. Regions may be based on physical characteristics, such as climate, soil type, or landforms. They may be based on cultural characteristics, such as political divisions, economy, religion, or language.

Latin America contains many regions. Five of these regions are listed in the box below. Decide which region or regions each Latin American country belongs to. Then write the letter of each region beside the country name. Remember, one country can belong to more than one region.

a. Central America
b. South America
c. Patagonia
d. the Andes
e. southern grasslands

___ 1. Argentina
___ 2. Belize
___ 3. Bolivia
___ 4. Brazil
___ 5. Chile
___ 6. Colombia
___ 7. Costa Rica
___ 8. Ecuador
___ 9. El Salvador
___10. French Guiana
___11. Guatemala
___12. Guyana
___13. Honduras
___14. Nicaragua
___15. Panama
___16. Paraguay
___17. Peru
___18. Suriname
___19. Uruguay
___20. Venezuela

© 2003 J. Weston Walch, Publisher

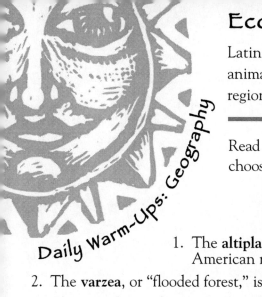

Ecosystems

Latin America includes a number of distinct **ecosystems**—plants and animals that make up an environment and affect each other. Each region has its own distinct characteristics including climate and terrain.

Read the descriptions of some of Latin America's ecosystems below. Then, choose the country or countries where each ecosystem can be found.

Argentina	Brazil	Mexico	Peru
Bolivia	Ecuador	Paraguay	Uruguay

1. The **altiplano** is a high plateau in the Andean regions of two Latin American nations.
2. The **varzea**, or "flooded forest," is found mainly along the Amazon River in this nation.
3. The **cerrado** is a dry savanna region found in South America's largest country.
4. The **Gran Chaco** is a flat plain with vegetation types including forest, savanna, and marsh.
5. The **pampas** is a huge savanna found east of the Andes in these two South American nations.
6. The **páramos** is a series of alpine meadows in this Andes nation.
7. The **selva** is a forested region found in three Latin American countries.
8. The **sertao** in this country is a region of thorny scrub forest where little rain falls.
9. The **zacaton** is a grassland region found only in the central part of this nation.

Absolute Location

Absolute location is the exact point on Earth where a place can be found. No two places have the same absolute location. This is usually found by using imaginary lines—latitude and longitude—that mark positions on the surface of the earth.

Match each place named in the left column with its latitude and longitude in the right column.

Place	Latitude and Longitude
____ 1. Bogotá, Colombia	a. 19° N, 99° W
____ 2. Havana, Cuba	b. 13° S, 73° W
____ 3. Machu Picchu, Peru	c. 5° N, 75° W
____ 4. Mexico City, Mexico	d. 23° S, 46° W
____ 5. São Paulo, Brazil	e. 23° N, 82° W

Relative Location

Every place on Earth has just one absolute location. This is the point on the latitude and longitude grid where it can be found. However, a place can also have a **relative location**. This means where it is relative to other places. If we say a place is north of, south of, or on the shores of, we are giving its relative location. For example, we might describe the relative location of São Paulo like this: "It is in southeastern Brazil, on the River Tiete, about 60 miles from the Atlantic Ocean."

For each place named below, write one or two sentences describing its relative location.

1. Belmopan, Belize
2. Buenos Aires, Argentina
3. Havana, Cuba
4. Lima, Peru
5. Rio de Janeiro, Brazil

Where Am I?

Use the clues in this paragraph to identify the place being described.

This landlocked country is southwest of Brazil and east of Peru and Chile. About a third of the country is on a high plateau in the Andes Mountains, the altiplano. There is snow on the highest mountain peaks—such as Nevado Sajama, which is 21,458 ft (6,542 m) high—throughout the year. Llamas, raised for leather and meat, graze high up in these mountains. One of the country's two capitals is found here, at an altitude of about 12,000 ft (4,000 m).

Lake Titicaca, the world's highest navigable lake, lies in the northwest of this country. Control of the lake is shared with Peru.

Toward the east, the land slopes down through tropical rain forests to the lowland plains of the Amazon Basin. Here, the weather is warm and humid. Tropical crops—such as coffee, cocoa, sugarcane, pineapples, banana, and avocados—grow well here.

In the southeast is a huge area of swamps and grasslands called the Gran Chaco. This region is rich in oil and natural gas—two of the country's most important exports. Other important exports include tin, zinc, silver, and gold.

Where am I?

© 2003 J. Weston Walch, Publisher

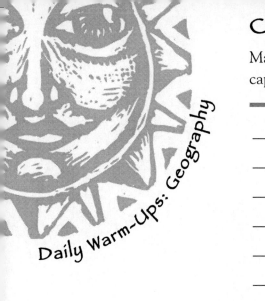

Capital Cities

Match each country name in the left column with the name of its capital city in the right column.

___ 1. Argentina   a. Asunción
___ 2. Brazil   b. Bogotá
___ 3. Chile   c. Brasília
___ 4. Colombia   d. Buenos Aires
___ 5. Ecuador   e. Caracas
___ 6. El Salvador f. Lima
___ 7. Mexico   g. Managua
___ 8. Nicaragua   h. Mexico City
___ 9. Panama   i. Panama City
___10. Paraguay   j. Quito
___11. Peru   k. Santiago
___12. Venezuela   l. San Salvador

Population of Mexico City

The graph below shows the population of Mexico City at different points in time. Study the graph carefully. Then answer the questions that follow.

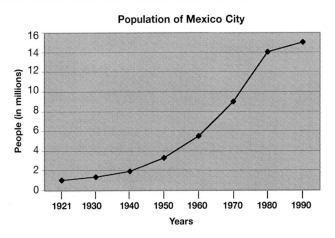

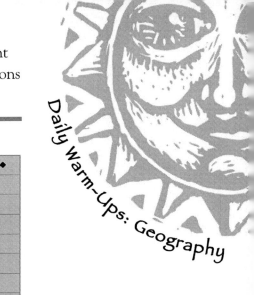

1. What was the approximate population of Mexico City in 1921?
2. What was the approximate population in 1990?
3. In which period did the population increase by the greatest number of people?
4. Nearly three quarters of the people in Latin America live in cities. What kinds of problems do you think this might cause?

© 2003 J. Weston Walch, Publisher

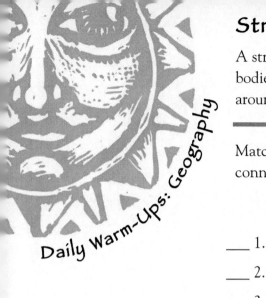

Strait and Narrow

A strait or channel is a narrow body of water that connects two larger bodies of water. There are approximately 200 straits or channels around the world. Some of them are in Latin American waters.

Match each strait in the left column with the bodies of water it connects, listed in the right column (some will be used twice).

Strait

___ 1. Panama Canal

___ 2. Strait of Magellan

___ 3. Drake Passage

___ 4. Dragon's Mouth and Serpent's Mouth

___ 5. Yucatán Channel

___ 6. Windward Passage

___ 7. Guadeloupe Passage

Bodies of Water

a. Gulf of Paria, Atlantic Ocean

b. Caribbean Sea, Atlantic Ocean

c. Atlantic Ocean, Pacific Ocean

d. Gulf of Mexico, Caribbean Sea

What's in a Name?

A **toponym** is a place name. Place names come from a variety of sources. Some places are named after people. Some are named for a distinctive physical feature. Some tell something about the culture of the place.

Each place name listed below means something. Match each name with its meaning. Use your knowledge of geography to help you find the answers.

____ 1. Argentina a. land of wood and water

____ 2. Costa Rica b. silver

____ 3. Ecuador c. land of many waters

____ 4. El Salvador d. where the land ends

____ 5. Guyana e. the savior

____ 6. Honduras f. little Venice

____ 7. Jamaica g. depths

____ 8. Paraguay h. equator

____ 9. Venezuela i. land of rivers

____ 10. Chile j. rich coast

Tropical Rain Forest

The tropical rain forests of South America contribute to the world in many ways. When rain falls, the forest soaks up huge volumes of water. Between rainstorms, the forest slowly releases water into rivers and streams. This provides water to other areas during times of drought. Also, trees absorb carbon dioxide from the atmosphere and produce oxygen. This helps all of us breathe purer air. Rain forests also contribute to species diversity. They contain almost one half of the earth's animal and plant species—some 30 million species in all.

The soil in most tropical forests is very thin. The high rainfall in the forest washes away any loose material on the ground. Because of this, dead plants or animals are quickly recycled by other plants and animals; nutrients are not stored in the soil, as they are in other types of forests.

Many people see other uses for the rain forest. Some want to harvest the trees to sell for wood. Some need land to plant crops. Both these groups cut down the trees that make up the forest. It is estimated that, every second, one and a half acres of rain forest trees are cut down.

Write a clear paragraph describing what happens to the environment when the rain forest trees are cut, either to harvest the trees or to clear fields for agriculture.

Multiple Choice

Choose the best answer for each question below.

---

1. Which Brazilian city, begun as a planned city in 1957, is located in a formerly uninhabited area of the country?

    a. Rio de Janeiro   b. Salvador   c. São Paulo   d. Brasília

2. What is the Mexican name for the Rio Grande?

    a. Rio Verde de Mato Grasso   c. Rio Conchos
    b. Rio Magdalena   d. Rio Bravo del Norte

3. Which two South American countries are completely landlocked?

    a. Bolivia and Paraguay   c. Paraguay and Uruguay
    b. Bolivia and Uruguay   d. Bolivia and Ecuador

4. Reggae, a kind of popular music that often addresses social and political issues, developed in the city of Kingston on which Caribbean island?

    a. Jamaica   b. Hispaniola   c. Puerto Rico   d. Cuba

5. Angel Falls, the world's highest waterfall at 3,212 ft (979 m), is located in which country?

    a. Argentina   b. Venezuela   c. Bolivia   d. Brazil

Unique Nations

Each feature described below is found in one of the countries listed in the box. Match each feature with the country where it is found. Some countries may include more than one feature.

| Bolivia | Mexico | Peru |
| Chile | Nicaragua | Venezuela |

_____ 1. The largest Spanish-speaking republic in the world

_____ 2. The world's largest bird colony

_____ 3. The world's tallest active volcano

_____ 4. The world's highest waterfall

_____ 5. The world's highest airport

_____ 6. The world's only freshwater sharks

_____ 7. The oldest capital city in the Americas

_____ 8. The world's longest nation

_____ 9. The world's driest place

_____ 10. The world's largest gulf

Daily Warm-Ups: Geography

© 2003 J. Weston Walch, Publisher

True or False?

Use your knowledge of the region to decide whether each of these statements is true or false. Write **T** for true or **F** for false on the line beside each statement.

_____ 1. There are no volcanoes in Mexico.

_____ 2. The Amazon has been thoroughly explored.

_____ 3. With a width of about 275 yds (250 m), the Monumental Axis in Brasília, Brazil, is the widest road in the world.

_____ 4. Puerto Rico is the most westerly of the Caribbean islands.

_____ 5. Nevados Ojos del Salado, in Chile, is the world's highest volcano.

_____ 6. The Antilles islands are separated into two groupings, the Westward and the Eastward.

_____ 7. Ski resorts in La Paz, Bolivia, only operate during the South American summer, as the winters are too cold.

_____ 8. All the countries in South America have a coastline on the Pacific, the Atlantic, or the Caribbean.

_____ 9. Costa Rica has barely 0.03% of the earth's land, but it is home to 6% of the world's animal and plant species, including more than 800 different types of birds.

_____ 10. The Andes form the longest mountain range on Earth.

Fill in the Blanks

Choose the correct word to complete each sentence below.

Argentina	Ecuador	Panama	Suriname
Brazil	Guyana	Peru	Venezuela
Colombia	Mexico		

1. Popocatepetl and Colima are the most active volcanoes in _____.

2. _____ is a thin sweep of land that connects Central America and South America.

3. The capital city of _____ was built to try to get people to move away from the coast.

4. The Falklands Islands, a U.K territory off the South American coast, are also claimed by _____.

5. The only country in South America to have a coast on both the Caribbean Sea and the Pacific Ocean is _____.

6. The _____ Current, also called the Humboldt Current, is a slow-moving stream of cold water in the southeast Pacific Ocean.

7. Oil from Lake Maracaibo makes up a large share of the economy of _____.

8. English is the main language in only one South American nation, _____.

Geography Terms

Different regions have different characteristics. Because of this, each region has its own terminology—words that are used to describe the region.

Each word below describes an aspect of the geography of Western Europe. Write a clear definition explaining each word.

bog

canton

dike

fjord

meseta

moor

peninsula

polder

Regions

Regions are groups of places that share at least one common characteristic. It may be physical (e.g., climate, soil type, landforms) or cultural (e.g., political divisions, economy, religion, language).

Five of the regions of Western Europe are listed in the box below. Decide which region or regions each country belongs to. Then write the letter of the correct region beside the country name. Remember, one country can belong to more than one region.

a. European Union
b. Iberia
c. Scandinavia
d. the Mediterranean
e. Romance language family

____ 1. Andorra
____ 2. Austria
____ 3. Belgium
____ 4. Denmark
____ 5. Finland
____ 6. France
____ 7. Germany
____ 8. Greece
____ 9. Iceland
____ 10. Ireland
____ 11. Italy
____ 12. Luxembourg
____ 13. Monaco
____ 14. the Netherlands
____ 15. Norway
____ 16. Portugal
____ 17. Spain
____ 18. Sweden
____ 19. Switzerland
____ 20. United Kingdom

Topography and Culture

Western Europe isn't very large; in area, it's about 428,000 sq mi (1,109,000 sq km). But this area is broken up into many smaller areas. Several parts of Europe consist of large islands. These include Great Britain, Ireland, and Iceland. Some are peninsulas, joined to the mainland by a comparatively narrow neck of land. These include Scandinavia, Iberia, and Italy. Even the mainland is broken up, crisscrossed by high mountain ranges.

Based on what you know about people and geography, how do you think the broken-up nature of Western Europe might have affected the way cultures developed here? Write one clear paragraph explaining your answer.

What's in a Name?

A **toponym** is a place name. Place names come from a variety of sources. Some places are named after people. Some are named for a distinctive physical feature. Some tell something about the culture of the place.

Each place name listed below means something. Match each name with its meaning. Use your knowledge of geography to help you find the answers.

____ 1. Dublin   a. the way to the north

____ 2. Mediterranean   b. lowlands

____ 3. Austria   c. in the middle of the earth

____ 4. Netherlands   d. black pool

____ 5. Norway   e. eastern kingdom

Absolute Location

Absolute location is the exact point on Earth where a place can be found. No two places have the same absolute location. This is usually found by using imaginary lines—latitude and longitude—that mark positions on the surface of the earth.

Match each place named in the left column with its latitude and longitude in the right column.

Place	Latitude and Longitude
____ 1. Athens, Greece	a. 60° N, 25° E
____ 2. Berlin, Germany	b. 53° N, 13° E
____ 3. Helsinki, Finland	c. 64° N, 22° W
____ 4. Lisbon, Portugal	d. 38° N, 24° E
____ 5. Reykjavik, Iceland	e. 39° N, 9° W

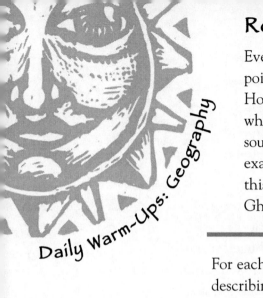

Relative Location

Every place on Earth has just one absolute location. This is the point on the latitude and longitude grid where it can be found. However, a place can also have a **relative location**. This means where it is relative to other places. If we say a place is north of, south of, or on the shores of, we are giving its relative location. For example, we might describe the relative location of Brussels like this: "It is in central Belgium, southwest of Antwerp, southeast of Ghent, and northwest of Waterloo."

For each place named below, write one or two sentences describing its relative location.

1. Hamburg, Germany
2. Kristianstad, Sweden
3. Marseille, France
4. Naples, Italy
5. Porto, Portugal

Where Am I?

Use the clues in this paragraph to identify the place being described.

This southwest European country occupies the western part of the Iberian peninsula. It is bounded by Spain to the north and east, and by the Atlantic Ocean to the south and west. Its capital city, Lisbon, is near the mouth of the Tagus River.

The Tagus River effectively divides the country in two. North of the river is a mountainous region. The climate here is influenced by air currents from the Atlantic Ocean and by the Meseta, Spain's huge central plateau. The area south of the river consists of lowlands, with a Mediterranean climate. There is little rainfall here, and temperatures tend to be warmer.

Because of its warm, sunny climate, the country's main agricultural products include citrus fruits, olives, and wine. It is also a major producer of cork, which comes from the bark of the cork oak. More than half the world's cork is produced here.

Where am I?

Climate

Much of Western Europe is at a fairly northerly latitude. For example, Ireland is located between latitudes 51° and 55° N. The United Kingdom lies between 50° and 60° N. Both these countries are at about the same latitude as Canada's Newfoundland and British Columbia.

Based just on latitude, these countries should have a subarctic climate, with cold winters and heavy snowfall. Instead, the climate is generally very mild. Ireland's average daytime temperature in December and January is 46° F (8° C). In July, it is 70° F (21° C). Even in Scotland, which is more mountainous and is cooled by the North Sea, the average temperature in January is around 40° F (4° C). At sea level, snow is rare.

Write two or three clear sentences explaining why the climate of the far west of Europe is so temperate.

I Got Sunshine

People often think that it is always cloudy in the United Kingdom. The graph below shows the average number of hours of sunshine each month in three locations in the United Kingdom: the southeast coast, the northeast coast, and in the central area. The fourth line on the chart shows the total number of hours of daylight each month, between sunrise and sunset.

Study the graph carefully. Then answer the questions that follow.

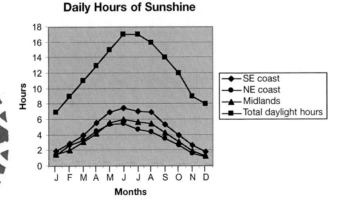

1. Which of the three locations receives the most sunshine?

2. Which receives the least sunshine?

3. The number of hours of daylight goes from a low of around 7 to a high of around 17. What do you think causes this wide variation?

Multiple Choice

Choose the correct answer for each question below.

1. If you were sailing on the Rhone River, which country would you be in?

    a. France   b. Belgium   c. Spain   d. Italy

2. Which of these Italian cities is farthest south?

    a. Florence   b. Rome   c. Naples   d. Pisa

3. Dijon mustard and Brie, a kind of cheese, are named for places in which European country?

    a. France   b. Spain   c. Italy   d. Austria

4. Paella, a dish made with rice, meat, seafood, and vegetables, is native to Valencia, a region in what Mediterranean country?

    a. France   b. Italy   c. Greece   d. Spain

5. Where are the Apennines in relation to the Alps?

    a. north   b. south   c. east   d. west

Capital Cities

Match each country name in the left column with the name of its capital city on the right.

___ 1. Austria   a. Berlin

___ 2. Denmark   b. Copenhagen

___ 3. France   c. Lisbon

___ 4. Germany   d. London

___ 5. Italy   e. Madrid

___ 6. Norway   f. Oslo

___ 7. Portugal   g. Paris

___ 8. Spain   h. Rome

___ 9. Sweden   i. Stockholm

___10. United Kingdom   j. Vienna

Peninsulas

A peninsula is a body of land that is surrounded by water on three sides. In Western Europe, three great peninsulas jut out into the Mediterranean Sea.

Name the four countries located on these peninsulas.

Germany—Together Again

Since the 1980s, the borders of many European countries have changed. One of these countries is Germany. After World War II, Germany was divided into two separate nations, East Germany and West Germany. Then, in 1990, the two nations were reunited. Germany is again one country.

Write a clear paragraph describing the challenges Germany had to face after it was reunited.

© 2003 J. Weston Walch, Publisher

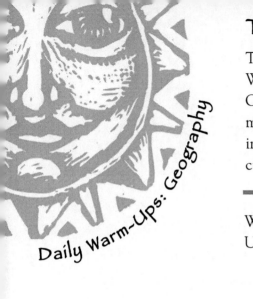

The European Union

The idea for a union of European nations grew out of World War II. It started small: Six nations formed the European Economic Community. Today, the European Union (EU) includes fifteen members. These countries have reduced barriers to doing business in Europe. As of 2002, twelve EU countries share a common currency, the euro.

Write a clear paragraph describing the advantages of the European Union for member nations.

Strait and Narrow

A strait or channel is a narrow body of water that connects two larger bodies of water. There are approximately 200 straits or channels around the world. Some of them are in Western European waters.

Match each strait in the left column with the bodies of water it connects, listed in the right column.

Strait

___ 1. North Channel
___ 2. Strait of Dover
___ 3. Skagerrak and Kattegat
___ 4. Strait of Gibraltar
___ 5. Strait of Sicily
___ 6. Strait of Otranto
___ 7. Bosporus
___ 8. Dardanelles
___ 9. Bonifacio Strait

Bodies of Water

a. Mediterranean Sea, Tyrrhenian Sea
b. Tyrrhenian Sea, Mediterranean Sea
c. Irish Sea, Atlantic Ocean
d. Adriatic Sea, Ionian Sea
e. Aegean Sea, Sea of Marmara
f. English Channel, North Sea
g. Atlantic Ocean, Mediterranean Sea
h. North Sea, Baltic Sea
i. Sea of Marmara, Black Sea

Unique Nations

Each numbered statement below describes one of the countries listed in the box. Match each statement with the country or countries it describes. Some countries may be used more than once.

Finland	the Netherlands	United Kingdom
France	Norway	Vatican City
Iceland	Spain	

1. The smallest country at less than one-fifth of a sq mi (.44 sq km) in area
2. The largest glacier in Europe, with an area of 182 sq mi (472 sq km), is in this country.
3. Although this country touches the Arctic Circle, houses here are heated by hot water from underground springs.
4. The lowest country in the world, with about 40% of the land below sea level
5. This country's language is not related to any other language in the world.
6. The country with the greatest number of islands: 179,584
7. The world's longest road tunnel, 31 mi (50 km) long (23 mi/37 km under water), connects these two countries.
8. The world's shortest intercontinental commercial flight, just 34 mi (55 km) long, flies to North Africa from here.

© 2003 J. Weston Walch, Publisher

Local Winds

In the study of weather, certain types of wind are expected for a given area. When winds do not fit this expected pattern, they are called **local winds**. That is, these winds do not come from the general pressure system for the region. They are usually caused by the effect on air masses of local topography—mountains, valleys, lakes, and so forth.

The winds described below are all local winds from one area: Lake Garda, in northern Italy. Look at the descriptions. Based on this information, what conclusions might you draw about the topography around Lake Garda? Write two or three clear sentences for your answer.

luganot: a strong south-to-southeast wind that blows off Lake Garda

ora: a strong southern wind that starts in the late morning along the shoreline of Lake Garda

paesa: a violent north-to-northeast wind that blows over Lake Garda

paesano: a night breeze from the north, blowing down from the mountains over Lake Garda

sarca: a strong north wind that blows over Lake Garda

tosca: a southwest wind that blows over Lake Garda

traersu: a violent east wind that blows over Lake Garda

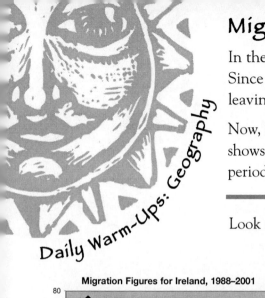

Migration Trends

In the mid-1800s, a famine forced millions of people to leave Ireland. Since then, the country has had a steady flow of emigrants—people leaving the country for a better life elsewhere.

Now, however, that pattern seems to be changing. The chart below shows figures for people leaving Ireland and moving to Ireland for the period 1988–2001.

Look at the chart, then answer the questions that follow.

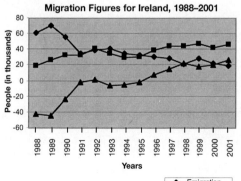

1. What changes does the chart show?
2. Based on what you know about people's reasons for emigration, what do these changes tell you about conditions in Ireland during this period?
3. Do you think a chart like this is a good way to show this information? Why or why not?

© 2003 J. Weston Walch, Publisher

Currency

In January 2002, 12 nations in Western Europe switched currency. Instead of using a separate currency for each nation, they now use one currency, the euro. However, some countries in Europe—including some in the European Union—have kept their own currencies. Match each country below with its currency in the box. (Some currencies are used in more than one country.) Then answer the question that follows.

| euro | krone | krona | pound | franc |

_____ 1. Austria

_____ 2. Belgium

_____ 3. Denmark

_____ 4. Finland

_____ 5. France

_____ 6. Germany

_____ 7. Greece

_____ 8. Iceland

_____ 9. Ireland

_____ 10. Italy

_____ 11. Luxembourg

_____ 12. Netherlands

_____ 13. Norway

_____ 14. Portugal

_____ 15. Spain

_____ 16. Sweden

_____ 17. Switzerland

_____ 18. United Kingdom

Why might some countries be willing to give up their own currencies for a shared currency? Write a clear paragraph for your answer.

True or False?

Use your knowledge of the region to decide whether each of these statements is true or false. Write **T** for true or **F** for false on the line beside each statement.

___ 1. There are no active volcanoes in the Mediterranean area.

___ 2. The language spoken in the Basque region of northern Spain is not related to any other language in the world.

___ 3. The weather in Greece, although warm, is usually overcast.

___ 4. The boundary between Belgium and the Netherlands is different on the surface and underground.

___ 5. About one fourth of Finland's electricity is generated by water power.

___ 6. Most of Western Europe is far removed from the sea.

___ 7. At 6,099 ft (1859 m), Mt. Snowden, in Wales, is the tallest mountain in Western Europe.

___ 8. The average elevation of the Netherlands is 1,000 ft (300 m) above sea level.

___ 9. Iceland has 200 active volcanoes; at least one erupts every five years.

___10. For several years in the 1800s, the capital of Portugal was located in Brazil, South America.

Fill in the Blanks

Choose the correct word to complete each sentence below.

Finland	Ireland	Switzerland
France	Italy	the United Kingdom
Germany	Sweden	

1. The Riksdag is the parliament of _____.
2. Much of the agriculture and industry of _____ is found in the Po River valley.
3. Because of its shape, _____ is sometimes called "the hexagon."
4. The landlocked, mountainous nation of _____ has remained neutral in all wars since 1815.
5. The Ruhr Valley is the main industrial region of _____.
6. The Giant's Causeway is a strange formation of rock columns in the north of _____.
7. The oil and gas fields of the North Sea are an important resource for the island nation of _____.
8. Squeezed in between Sweden and Russia, _____ has built a prosperous economy on natural resources such as wood and water.

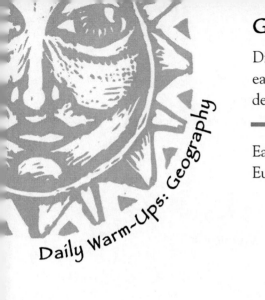

Geography Terms

Different regions have different characteristics. Because of this, each region has its own terminology—words that are used to describe the region.

Each word below describes an aspect of the geography of Eastern Europe. Write a clear definition for each word.

    bora

    chernozem

    karst

    loess

    puszta

    rendzina

Absolute Location

Absolute location is the exact point on Earth where a place can be found. No two places have the same absolute location. This is usually found by using imaginary lines—latitude and longitude—that mark positions on the surface of the earth.

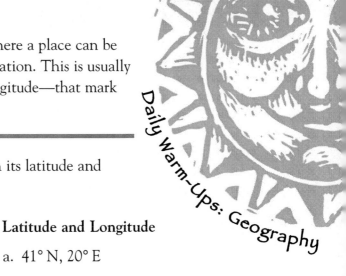

Match each place named in the left column with its latitude and longitude in the right column.

Place	Latitude and Longitude
____ 1. Belgrade, Serbia	a. 41° N, 20° E
____ 2. Bucharest, Romania	b. 45° N, 21° E
____ 3. Ljubljana, Slovenia	c. 44° N, 26° E
____ 4. Tiranë, Albania	d. 52° N, 21° E
____ 5. Warsaw, Poland	e. 46° N, 15° E

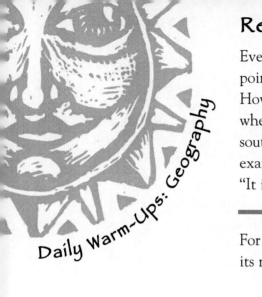

Daily Warm-Ups: Geography

Relative Location

Every place on Earth has just one absolute location. This is the point on the latitude and longitude grid where it can be found. However, a place can also have a **relative location**. This means where it is relative to other places. If we say a place is north of, south of, or on the shores of, we are giving its relative location. For example, we might describe the relative location of Prague like this: "It is in the north central Czech Republic, on the Vltava River."

For each place named below, write one or two sentences describing its relative location.

1. Brno, Czech Republic
2. Bucharest, Romania
3. Durres, Albania
4. Kraków, Poland
5. Maribor, Slovenia

© 2003 J. Weston Walch, Publisher

Where Am I?

Use the clues in this paragraph to identify the place being described.

Located in central Europe, northwest of Romania, this landlocked country is about 36,000 sq mi (93,030 sq km) in area. Neighboring countries include Austria, Croatia, Romania, Serbia and Montenegro, Slovakia, Slovenia, and Ukraine.

Most of the land in this country consists of flat or rolling plains, with hills and low mountains to the north. The Danube River runs from Slovakia in the north to Yugoslavia in the south, through the capital city. The other major river in the country is the Tisza, which flows from the northeast to the south.

The country's main resource is the rich farmland of the plains. The temperate climate, with cold, cloudy, humid winters and warm summers, means that corn, wheat, potatoes, and sugar beets grow well here.

Although this country is landlocked, historically, its location has been strategic. It lies across the main land routes between Western Europe and the Balkan Peninsula, as well as routes between Ukraine and the Mediterranean.

Where am I?

Country Car Codes

When travelers drive between different countries, they usually need to put an identifying sticker on their car. The sticker shows the country that the car comes from. These stickers are usually a white oval with a letter or letters for the country of origin.

These letters have been specified for each country, to prevent confusion. However, some of these codes are based on the language spoken in the country, not on English. For example, the country we know as "Germany" is actually "Deutschland" in German. Its country code is "D."

Below are the country codes for several countries in Eastern Europe. Choose the correct country name for each country code sticker.

Albania	Bulgaria	Slovakia	Macedonia
Bosnia/Herzegovina	Hungary	Croatia	Czech Republic
Serbia and Montenegro	Poland	Romania	Slovenia

_____ 1. AL   _____ 4. MK   _____ 7. SCG   _____ 10. H

_____ 2. BG   _____ 5. RO   _____ 8. SK   _____ 11. HR

_____ 3. CZ   _____ 6. SLO _____ 9. PL   _____ 12. BiH

© 2003 J. Weston Walch, Publisher

The Danube River

The Danube, the second-longest river in Europe, flows through much of Eastern Europe. It rises in Germany, then flows about 1,770 mi (2,850 km) through Austria, Slovakia, Hungary, Serbia, Croatia, Bosnia and Herzegovina, Slovenia, Bulgaria, Romania, and Ukraine. The Eastern European cities of Belgrade, Braila, Bratislava, Budapest, and Galati are all on the Danube. The river is the subject of a famous waltz, "The Blue Danube."

What factors make the Danube so important? Write a clear paragraph about the river, explaining its importance in Eastern Europe.

Multiple Choice

Choose the correct answer for each question below.

1. In which country would you find the city of Gdansk?
    a. Bulgaria   b. Poland   c. Romania   d. Ukraine
2. Of which country is Ljubljana the capital?
    a. Moldova   b. Slovakia   c. Latvia   d. Slovenia
3. The Danube River forms most of the border between which two countries?
    a. Austria and Germany   c. Austria and Hungary
    b. Bulgaria and Romania   d. Croatia and Slovenia
4. More than 90 percent of _____, the capital of Poland, was destroyed in World War II, but the city's historic Old Town section was painstakingly restored, and is now a financial and cultural center of eastern Europe.
    a. Warsaw   b. Lódz   c. Kraków   d. Gdansk
5. Where are the Carpathian Mountains?
    a. Croatia, Poland, and Yugoslavia   c. Slovakia, Ukraine, and Romania
    b. Albania, Bulgaria, and Slovakia   d. Romania, Hungary, and Poland

Per Capita Income

Per capita means "for each head." Per capita income is a measure of a country's wealth. It shows how much each person in the country would earn if the country's total income was divided by the number of inhabitants. Of course, some people in a country earn much more than the per capita income. And some earn much less.

Study the graph carefully. Then answer the questions that follow.

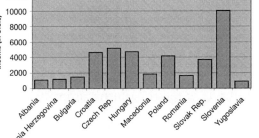

1. Is the per capita income roughly the same in all the countries in this region, or is there a wide variation?
2. About what is the per capita income in the Czech Republic?
3. About what is the per capita income in Slovenia?
4. What is the difference in per capita income between the Czech Republic and Slovenia?

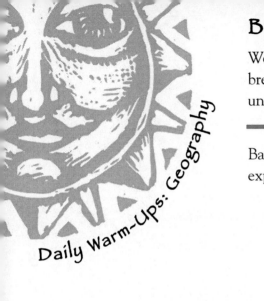

Balkanize

Webster's dictionary defines the word *balkanize* as meaning "to break up (as a region or group) into smaller and often hostile units." This word comes from the Balkan region of Eastern Europe.

Based on your knowledge of the region, write one or two sentences explaining the origin of the word *balkanize*.

Local Winds

In the study of weather, certain types of wind are expected for a given area. When winds do not fit this expected pattern, they are called **local winds**. That is, these winds do not come from the general pressure system for the region.

The winds described below are all local winds in parts of Eastern Europe. Look at the descriptions. Do you see any patterns? Based on this information, what aspects of geography do you think help cause local winds?

bora: a cold, gusty wind from the north or northeast on the eastern shore of the Adriatic Sea

karajol: a west wind that follows rain on the Bulgarian coast

kossava: a cold wind that descends from the east or southeast in the area of the Danube pass through the Carpathians

lodos: a southerly wind that occurs on the Black Sea coast of Bulgaria

maestro: a northwesterly wind that brings clear weather to the Adriatic region

polacke: a cold, dry, northeasterly wind caused by air descending from the Sudeten mountains

Capital Cities

Match each country name in the left column with the name of its capital city on the right.

____ 1. Albania   a. Belgrade

____ 2. Bosnia and Herzegovina   b. Bucharest

____ 3. Bulgaria   c. Budapest

____ 4. Croatia   d. Prague

____ 5. Czech Republic   e. Sarajevo

____ 6. Hungary   f. Skopje

____ 7. Macedonia   g. Sofia

____ 8. Poland   h. Tiranë

____ 9. Romania   i. Warsaw

____ 10. Serbia and Montenegro   j. Zagreb

Currency

Match each country named below with its currency, listed in the box. Some currencies are used in more than one country.

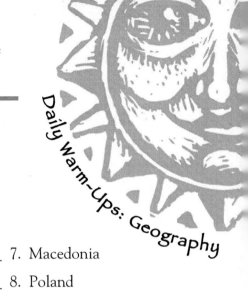

denar	koruna	leu	tolar
dinar	kuna	leva	zloty
forint	lek	marka	

_____ 1. Albania

_____ 2. Bosnia and Herzegovina

_____ 3. Bulgaria

_____ 4. Croatia

_____ 5. Czech Republic

_____ 6. Hungary

_____ 7. Macedonia

_____ 8. Poland

_____ 9. Romania

_____ 10. Slovakia

_____ 11. Slovenia

_____ 12. Serbia and Montenegro

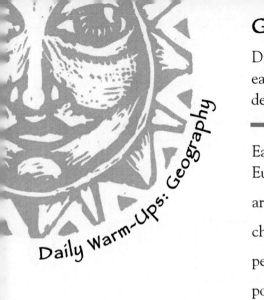

Geography Terms

Different regions have different characteristics. Because of this, each region has its own terminology—words that are used to describe the region.

Each word below describes an aspect of the geography of Northern Eurasia. Write a clear definition explaining each word.

arctic

chernozem

permafrost

podzol

steppe

taiga

tundra

Regions

Regions are groups of places that share at least one common characteristic. It may be physical (e.g., climate, soil type, landforms) or cultural (e.g., political divisions, economy, religion, language).

Four of the regions of Northern Eurasia are listed in the box below. Decide which region or regions each country belongs to. Then write the letter of the correct region beside the country name.

a. Baltic States	c. Commonwealth of Independent States (CIS)
b. Central Asia	d. Transcaucasia

___ 1. Armenia

___ 2. Azerbaijan

___ 3. Belarus

___ 4. Estonia

___ 5. Georgia

___ 6. Kazakhstan

___ 7. Kyrgyzstan

___ 8. Latvia

___ 9. Lithuania

___ 10. Moldova

___ 11. Russia

___ 12. Tajikistan

___ 13. Turkmenistan

___ 14. Ukraine

___ 15. Uzbekistan

© 2003 J. Weston Walch, Publisher

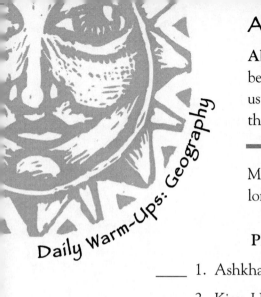

Absolute Location

Absolute location is the exact point on Earth where a place can be found. No two places have the same absolute location. This is usually found by using imaginary lines—latitude and longitude—that mark positions on the surface of the earth.

Match each place named in the left column with its latitude and longitude in the right column.

Place	Latitude and Longitude
____ 1. Ashkhabad, Turkmenistan	a. 59° N, 25° E
____ 2. Kiev, Ukraine	b. 40° N, 44° E
____ 3. Moscow, Russia	c. 38° N, 58° E
____ 4. Tallinn, Estonia	d. 55° N, 37° E
____ 5. Yerevan, Armenia	e. 50° N, 31° E

Relative Location

Every place on Earth has just one absolute location. This is the point on the latitude and longitude grid where it can be found. However, a place can also have a **relative location**. This means where it is relative to other places. If we say a place is north of, south of, or on the shores of, we are giving its relative location. For example, we might describe the relative location of Archangel like this: "It is in western Russia, on the White Sea, northeast of Moscow, northwest of Novgorod."

For each place named below, write one or two sentences describing its relative location.

1. Almaty, Kazakhstan
2. Gomel, Belarus
3. Gyumri, Armenia
4. St. Petersburg, Russia
5. Yalta, Ukraine

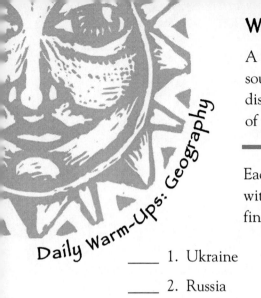

What's in a Name?

A **toponym** is a place name. Place names come from a variety of sources. Some places are named after people. Some are named for a distinctive physical feature. Some tell something about the culture of the place.

Each place name listed below means something. Match each name with its meaning. Use your knowledge of geography to help you find the answers.

____ 1. Ukraine   a. land of free people

____ 2. Russia   b. country of those wearing crowns

____ 3. Azerbaijan   c. land of fire

____ 4. Kazakhstan   d. border or frontier

____ 5. Kyrgyzstan   e. land of genuine man

____ 6. Tajikistan   f. land of 40 tribes

____ 7. Uzbekistan   g. men of the rowing way

© 2003 J. Weston Walch, Publisher

Where Am I?

Use the clues in this paragraph to identify the place being described.

This landlocked nation lies between Russia to the east, Poland to the west, Ukraine to the south, and Lithuania to the north. Most of the country is one large plain, broken by a range of low-lying hills.

It has a transitional climate, between continental and maritime. Its winters are cold and its summers are cool and moist. Combined with the country's flat terrain, this makes agriculture an important part of the economy. Agricultural products include grain, potatoes, vegetables, sugar beets, flax, beef, and milk. Other natural resources include forests (over one third of the country is forested), peat, and small quantities of oil and natural gas.

In 1939, this country was annexed by Soviet military forces. From then until December 1991, it was a republic within the Soviet Union. It is now a member of the Commonwealth of Independent States.

Where am I?

Cars in Estonia

Estonia lies on the Gulf of Finland and the Baltic Sea, to the east of Russia. It is a small country, with a population of 1.4 million people and about 17,000 sq mi (45,000 sq km) of land. Much of the mainland is boggy and partly wooded. The country also includes more than 1,500 offshore islands.

Before the fall of the Soviet Union, Estonia was part of the Communist economy. Few Estonians owned private cars. Out of every 1,000 people, about 150 owned a car. Since 1991, private car ownership has increased. By 1993, 211 people in 1,000 owned a car. By 1997, the figure was 294 in 1,000. By 2010, it is estimated that 385 people in 1,000 will own a car.

This increase in car ownership tells a lot about social and political changes since the Soviet Union broke up. However, this increase will also affect Estonia's environment and infrastructure. Using your knowledge about Estonia's geography, write two or three clear sentences describing the kinds of effects Estonia might face.

Trans-Siberian Railway

The world's longest continuous rail line runs from Moscow to Vladivostok. It covers almost 6,000 mi (10,000 km), and passes through seven time zones. Nonstop, it takes 153 hours and 49 minutes, or $6\frac{1}{2}$ days! Even though the train crosses seven time zones, the train timetable is based on Moscow time.

Part of the timetable for the Trans-Siberian Railway is shown below. Look at it closely. Then answer the questions that follow.

Timetable from Moscow to Vladivostok

Total km traveled	City	Time (Moscow Time)			Day	Time zone
		Arr.	Stop	Dep.		
0	Moscow			15:26	1	0
957	Kirov	05:28	15	5:43	2	+ 1
2716	Omsk	06:48	25	07:13	3	+ 3
5192	Irkutsk	21:25	20	21:45	4	+ 5
9298	Vladivostok	23:53			7	+ 7

1. If you set your watch forward one hour each time you changed time zones, what time would your watch say as you pulled out of Omsk?

2. What would be the local time when you arrived in Vladivostok?

3. If you left Moscow on June 11, what day would it be in Vladivostok when you arrived?

Ukraine, the Breadbasket

After Russia, Ukraine is the largest country in Eurasia. To the north and east, it is bordered by Russia and Belarus. On the west lie Poland, Slovakia, Hungary, Romania, and Moldova. The Black Sea and the Sea of Azov are to the south.

This area was once known as the "Breadbasket of Europe" and the "Breadbasket of the U.S.S.R." Based on what you know about the region, write two or three clear sentences explaining these nicknames.

The Cyrillic Alphabet

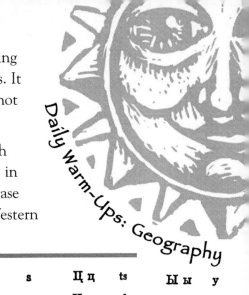

Several languages of Eurasia use the Cyrillic alphabet. This writing system has symbols for the distinctive sounds of Slavic languages. It uses 33 letters: 21 consonants, 10 vowels, and 2 letters that are not sounded.

The table below shows some letters of the Cyrillic alphabet, with the letters of the Western alphabet used to show Russian sounds in English. Study the table. Then use it to match each Russian phrase written in Cyrillic below with the same phrase written in the Western alphabet.

А а	a	Ж ж	zh	М м	m	С с	s	Ц ц	ts	Ы ы	y
Б б	b	З з	z	Н н	n	Т т	t	Ч ч	ch	Ь ь[10]	,
В в	v	И и Й й	i, ĭ	О о	o	У у	u	Ш ш	sh	Э э	e
Г г	g	К к	k	П п	p	Ф ф	f	Щ щ	shch	Ю ю	yu
Д д	d	Л л	l	Р р	r	Х х	kh	Ъ ъ[9]	"	Я я	ya
Е е	e										

____ 1. Dobry dyen (Good day)

____ 2. Do svidaniya (Good-bye)

____ 3. Spasibo (Thank-you)

a. СПАСИБО

b. ДОБРЫ ДЫЕН

c. ДО СВИДАНИЯ

Siberian Tiger

The Siberian tiger lives in the mixed forests of eastern Russia. Over the last 100 years, large areas of the forest have been cut down, reducing tiger numbers.

Tigers are also killed by poachers. When a law was passed to ban tiger hunting, tiger numbers began to recover. When the Soviet Union broke up, poaching increased.

Russia has now said that the tiger is one of the country's most valuable natural objects, and has set up a conservation program.

The graph below shows approximate numbers of Siberian tigers at different times. Study the graph carefully. Then answer the questions that follow.

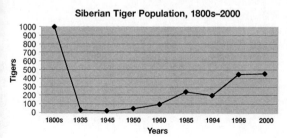

1. Based on the graph, when do you think the law against tiger hunting was passed?

2. When do you think poaching tigers again became a problem?

3. When do you think the anti-poaching measures went into effect?

Capital Cities

Match each country name in the left column with the name of its capital city on the right.

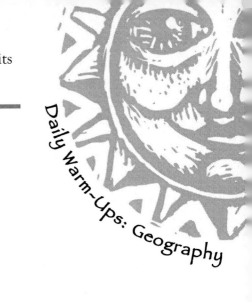

____ 1. Azerbaijan   a. Baku

____ 2. Belarus   b. Dushanbe

____ 3. Estonia   c. Kiev

____ 4. Georgia   d. Minsk

____ 5. Latvia   e. Moscow

____ 6. Lithuania   f. Riga

____ 7. Russia   g. Tallinn

____ 8. Tajikistan   h. Tashkent

____ 9. Ukraine   i. Tbilisi

____ 10. Uzbekistan   j. Vilnius

© 2003 J. Weston Walch, Publisher

Currency

Match each country named below with its currency, listed in the box. Some currencies are used in more than one country.

dram	lari	litas	som	tenge
hryvnia	lat	manat	somoni	
kroon	leu	ruble	sum	

_____ 1. Armenia

_____ 2. Azerbaijan

_____ 3. Belarus

_____ 4. Estonia

_____ 5. Georgia

_____ 6. Kazakhstan

_____ 7. Kyrgyzstan

_____ 8. Latvia

_____ 9. Lithuania

_____ 10. Moldova

_____ 11. Russia

_____ 12. Tajikistan

_____ 13. Turkmenistan

_____ 14. Ukraine

_____ 15. Uzbekistan

© 2003 J. Weston Walch, Publisher

St. Petersburg

St. Petersburg is in Russia's far west, on the Baltic Sea. It was founded in 1703 by Peter the Great, the czar of Russia. Peter fought several wars to get control of this area. Until then, all of Russia's ports were in arctic waters. They could only be used in summer. In winter, the harbors froze and ships could no longer enter the ports.

Think about what you know about Russia's geography and what transportation was like in the 1700s. Then write a clear paragraph explaining why having a port on the Baltic Sea was so important to the czar.

Bonus question: Over the years, St. Petersburg has had several different names. What were they? Why were the name changes made?

True or False

Use your knowledge of the region to decide whether each of these statements is true or false. Write **T** for true or **F** for false on the line beside each statement.

___ 1. In terms of area, Russia is the largest country in the world.

___ 2. Very little of Russia's land is forested.

___ 3. With its mild, dry climate, the Baltic Sea area is a popular Russian vacation spot.

___ 4. All of Georgia is in Asia, not in Europe.

___ 5. In Siberia, it can get so cold that the moisture in breath freezes. It can be heard when it falls to the ground as ice crystals.

___ 6. Murmansk, in Russia, is the largest city north of the Arctic Circle.

___ 7. Lake Baikal in Siberia is the only lake in the world that is deep enough to have deep-sea fish.

___ 8. All the countries in this region have a long history of independence.

___ 9. Most of this region is very densely populated.

___10. The countries in this region are united by common languages and ethnic backgrounds.

Geography Terms

Different regions have different characteristics. Because of this, each region has its own terminology—words that are used to describe the region.

Each word below describes an aspect of the geography of Southwest Asia. Write a clear definition for each word.

desert

irrigation

monsoon

oasis

wadi

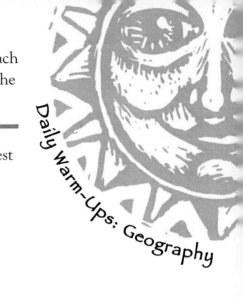

Absolute Location

Absolute location is the exact point on Earth where a place can be found. No two places have the same absolute location. This is usually found by using imaginary lines—latitude and longitude—that mark positions on the surface of the earth.

Match each place named in the left column with its latitude and longitude in the right column.

Place	Latitude and Longitude
____ 1. Amman, Jordan	a. 15° N, 44° E
____ 2. Ankara, Turkey	b. 36° N, 52° E
____ 3. Muscat, Oman	c. 24° N, 59° E
____ 4. Sanaa, Yemen	d. 32° N, 36° E
____ 5. Tehran, Iran	e. 40° N, 33° E

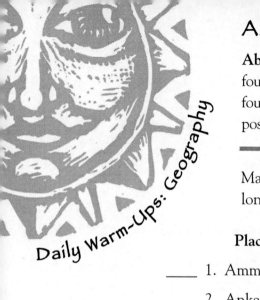

Relative Location

Every place on Earth has just one absolute location. This is the point on the latitude and longitude grid where it can be found. However, a place can also have a **relative location**. This means where it is relative to other places. If we say a place is north of, south of, or on the shores of, we are giving its relative location. For example, we might describe the relative location of Baghdad like this: "It is in central Iraq, on the Tigris River, to the south of the fertile plains between the Tigris and the Euphrates."

For each place named below, write one or two sentences describing its relative location.

1. Abu Dhabi, United Arab Emirates
2. Beirut, Lebanon
3. Kuwait, Kuwait
4. Manama, Bahrain
5. Tel Aviv, Israel

Where Am I?

Use the clues in this paragraph to identify the place being described.

This country was once known as the Ottoman Empire. Roughly oval in shape, it stretches from Asia to Europe. In fact, the largest city is located half in Asia and half in Europe, with the straits of the Bosporus flowing down the center. To the north is the Black Sea, to the west the Aegean, and to the southwest the Mediterranean. In Europe this country borders Greece and Bulgaria. In Asia, it borders Syria, Iraq, Iran, Armenia, Azerbaijan, and Georgia.

A narrow coastal plain provides fields for crops like wheat and corn, and for olive groves, orange trees, and cotton. Away from the coast, the land slopes upward. A huge plateau covers the center of the country; summers here are hot, and winters very cold. East of the plateau, the land slopes up again to the Taurus and Pontic mountains. The highest peak is Mt. Ararat, near the borders of Armenia, Azerbaijan, and Iran.

The country is subject to severe earthquakes, especially in the north. In recent years, several earthquakes have caused great destruction.

Where am I?

Capital Cities

Match each country name on the left with the name of its capital city on the right.

____	1. Iran	a. Abu Dhabi
____	2. Iraq	b. Amman
____	3. Israel	c. Baghdad
____	4. Jordan	d. Beirut
____	5. Lebanon	e. Damascus
____	6. Oman	f. Jerusalem
____	7. Saudi Arabia	g. Muscat
____	8. Syria	h. Riyadh
____	9. United Arab Emirates	i. Sanaa
____	10. Yemen	j. Tehran

Daily Warm-Ups: Geography

118

© 2003 J. Weston Walch, Publisher

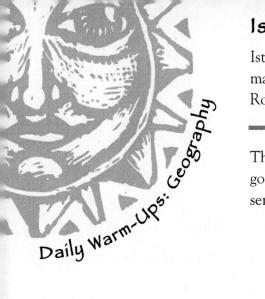

Istanbul

Istanbul. Constantinople. Stamboul. The city of Istanbul has had many names and many rulers. It was the center of the Eastern Roman Empire, as well as of the Ottoman Empire.

Think about Istanbul's location. Why would this make the city a good place from which to run an empire? Write one or two clear sentences to explain your answer.

Strait and Narrow

A strait or channel is a narrow body of water that connects two larger bodies of water. There are approximately 200 straits or channels around the world. Some of them are in Southwest Asian waters.

Match each strait in the left column with the bodies of water it connects, listed in the right column.

Strait

____ 1. Bosporus

____ 2. Dardanelles

____ 3. Suez Canal

____ 4. Strait of Hormuz

____ 5. Bab el Mandeb

____ 6. Strait of Tiran

Bodies of Water

a. Mediterranean Sea, Red Sea

b. Red Sea, Gulf of Aden

c. Aegean Sea, Sea of Marmara

d. Gulf of Aqaba, Red Sea

e. Sea of Marmara, Black Sea

f. Persian Gulf, Gulf of Oman

The Arabian Peninsula

The Arabian Peninsula is a vast wedge of land between the Red Sea, the Gulf of Aden, the Gulf of Oman, and the Persian Gulf.

Name the seven nations located on this peninsula.

Population

The chart below gives some information about populations in countries of Southwest Asia. Study the chart carefully. Then answer the questions that follow.

Country	Population mid-2001 (millions)	Births per 1,000 People	Deaths per 1,000 People	Rate of Natural Increase (%)
Bahrain	0.7	21	3	1.9
Cyprus	0.9	13	8	0.6
Iran	66.1	18	6	1.2
Iraq	23.6	37	10	2.7
Israel	6.4	22	6	1.6
Jordan	5.2	27	5	2.2
Lebanon	4.3	23	7	1.7
Oman	2.4	39	4	3.5
Saudi Arabia	21.1	35	6	2.9
Turkey	66.3	22	7	1.5

1. What is the general trend for populations in this region?

2. Which country has the greatest percentage of increase?

3. Which country will have the greatest increase in the number of people?

Currency

Match each country named below with its currency, listed in the box. Some currencies are used in more than one country.

afghani	lira	rial
dinar	new shekel	riyal
dirham	pound	

_____ 1. Afghanistan
_____ 2. Bahrain
_____ 3. Cyprus
_____ 4. Iran
_____ 5. Iraq
_____ 6. Israel
_____ 7. Jordan
_____ 8. Kuwait
_____ 9. Lebanon
_____ 10. Libya
_____ 11. Oman
_____ 12. Qatar
_____ 13. Saudi Arabia
_____ 14. Syria
_____ 15. Turkey
_____ 16. United Arab Emirates
_____ 17. Yemen

Daily Warm-Ups: Geography

123

The Suez Canal

The Suez Canal connects the Mediterranean Sea and the Red Sea. Below are some of the key dates in the history of the Suez Canal. Use them to create a time line of the canal from the time it was first thought of until the time the modern canal was opened. Then answer the question that follows.

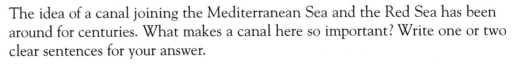

c. 1300 B.C.E.	A canal is built between the Nile delta and the Red Sea.
1300 B.C.E.–800 C.E.	The canal is intermittently maintained and neglected.
c. 110 C.E.	Canal redug under Roman emperor Trajan.
c. 799 C.E.	Canal redug by Arab ruler Amr Ibn-Al-Aas.
c. 1500 C.E.	Canal abandoned.
1854	Muhammad Said, viceroy of Egypt, agrees to canal plan devised by Ferdinand de Lesseps.
1855	The Mixed Commission studies proposed canal route.
1859	Despite opposition, work on canal begins.
1869	Suez Canal opens.

The idea of a canal joining the Mediterranean Sea and the Red Sea has been around for centuries. What makes a canal here so important? Write one or two clear sentences for your answer.

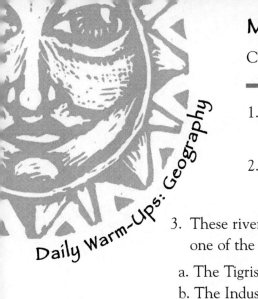

Multiple Choice

Choose the correct answer for each question below.

---

1. Mecca is located in this country.

    a. Saudi Arabia   b. Kuwait   c. Iran   d. Syria

2. This is the official language of Kuwait.

    a. Urdu   b. Persian (Farsi)   c. Kurdish   d. Arabic

3. These rivers, which flow through Turkey, Syria, and Iraq, were the site of one of the world's earliest civilizations.

    a. The Tigris and the Nile   c. The Tigris and the Euphrates
    b. The Indus and the Euphrates   d. The Indus and the Nile

4. The world's lowest body of water is located between these two countries.

    a. Israel and Jordan   c. Jordan and Iraq
    b. Israel and Syria   d. Syria and Iraq

5. This city, founded around 6000 B.C.E., is the oldest continuously inhabited city in existence.

    a. Beirut, Lebanon   b. Baghdad, Iraq   c. Damascus, Syria   d. Jerusalem, Israel

© 2003 J. Weston Walch, Publisher

Geography Terms

Different regions have different characteristics. Because of this, each region has its own terminology—words that are used to describe the region.

Each word below describes an aspect of the geography of Africa. Write a clear definition of each.

cataract

delta

desert

equator

harmattan

sahel

savanna

sirocco

veldt

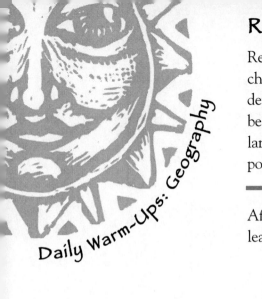

Regions

Regions are groups of places that have at least one common characteristic. One area may be part of many different regions, depending on which characteristic is being considered. Regions may be based on physical characteristics, such as climate, soil type, or landforms. They may be based on cultural characteristics, such as political divisions, economy, religion, or language.

Africa contains many regions; five of them are listed below. List at least three African countries for each region.

North Africa

West Africa

East Africa

Central Africa

Southern Africa

Bar Chart

Look at the bar chart below. What does it show? How can you tell?

Write a descriptive label for the chart. Then write two or three clear sentences explaining what the chart shows.

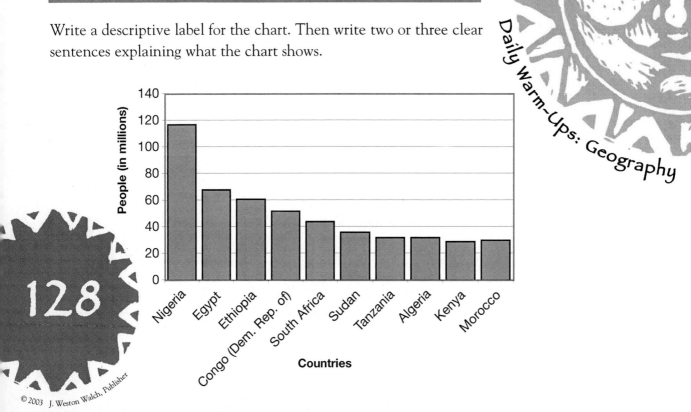

Daily Warm-Ups: Geography

Unique Nations

Each feature described below is found in one of the countries listed in the box. Match each feature with the country where it is found. Some countries may include more than one feature.

| Egypt | Libya | South Africa | Tanzania |
| Liberia | Namibia | Sudan | |

_____ 1. The only desert inhabited by elephant, rhino, giraffe, and lion

_____ 2. The world's longest river

_____ 3. The only country outside the United States with a capital named for a U.S. president

_____ 4. The hottest place in the world

_____ 5. The largest country in Africa

_____ 6. The highest waterfall in Africa

_____ 7. The driest inhabited region in the world

_____ 8. The highest mountain in Africa

_____ 9. The largest nonpolar desert in the world

129

© 2003 J. Weston Walch, Publisher

What's in a Name?

A **toponym** is a place name. Place names come from a variety of sources. Some places are named after people. Some are named for a distinctive physical feature. Some tell something about the culture of the place.

Each place name listed below means something. Match each name with its meaning. Use your knowledge of geography to help you find the answers.

____ 1. Algeria   a. house of the spirit
____ 2. Burkina Faso   b. area where there is nothing
____ 3. Côte d'Ivoire   c. land of honest men
____ 4. Egypt   d. land of the Moors
____ 5. Kenya   e. great stone building
____ 6. Liberia   f. the island
____ 7. Mauritania   g. free land
____ 8. Namibia   h. ivory coast
____ 9. Sierra Leone   i. lion mountains
____10. Zimbabwe   j. mountain of whiteness

Daily Warm-Ups: Geography

Food Shortages

Some African countries south of the Sahara cannot produce enough food every year. In 2002, the United Nations said that about 20 million people in 16 countries faced food shortages.

In some countries, this is the result of drought. Sometimes there is not enough rain to grow crops. In other areas, war creates food shortages. People are forced to leave their homes to avoid fighting. This means they cannot cultivate the land. It also means that these people end up in another area, putting new demands on food resources there.

Food shortages are particularly hard on children. Their small bodies can't go without food for long.

The short-term solution to these problems is to provide food for people who need it. Organizations such as the United Nations World Food Programme (WFP) try to do this. However, this is not a long-term solution.

How do you think the underlying causes of hunger in sub-Saharan Africa can be solved? Write down as many possible solutions as you can.

Animals of the Desert

During the day, the desert may seem like a very quiet place. There are few animals around. The landscape is a bright blur of sun and sand.

However, when the sun sets, the desert comes alive. Many desert animals are **crepuscular**. That is, they are only active at dusk and at dawn. Others are **nocturnal**. They sleep during the day and are active at night. People who cross the desert by day may never see these animals.

Why do you think desert animals are not active during the day? Write two or three clear sentences explaining why they might have evolved to be nocturnal or crepuscular.

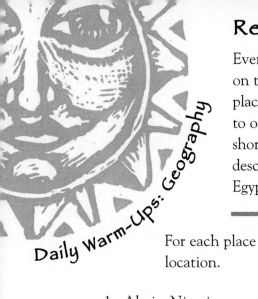

Relative Location

Every place on Earth has just one absolute location. This is the point on the latitude and longitude grid where it can be found. However, a place can also have a **relative location**. This means where it is relative to other places. If we say a place is north of, south of, or on the shores of, we are giving its relative location. For example, we might describe the relative location of Cairo like this: "It is in northeastern Egypt, southeast of Alexandria, northwest of Aswan."

Daily Warm-Ups: Geography

For each place listed below, write one or two sentences describing its relative location.

1. Abuja, Nigeria
2. Antananarivo, Madagascar
3. Bamako, Mali
4. Brazzaville, Republic of the Congo
5. Dodoma, Tanzania
6. Khartoum, Sudan
7. Kigali, Rwanda
8. Monrovia, Liberia
9. Niamey, Niger
10. Tunis, Tunisia

© 2003 J. Weston Walch, Publisher

Capital Cities

Match each country name on the left with the name of its capital city on the right.

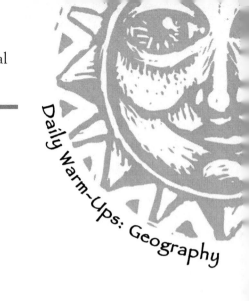

____ 1. Algeria   a. Accra

____ 2. Angola   b. Algiers

____ 3. Burundi   c. Asmara

____ 4. Eritrea   d. Bujumbura

____ 5. Gabon   e. Kampala

____ 6. Ghana   f. Libreville

____ 7. Libya   g. Lilongwe

____ 8. Malawi   h. Luanda

____ 9. Mozambique   i. Maputo

____ 10. Uganda   j. Tripoli

Absolute Location

Absolute location is the exact point on Earth where a place can be found. No two places have the same absolute location. This is usually found by using imaginary lines—latitude and longitude—that mark positions on the surface of the earth.

Match each place named in the left column with its latitude and longitude in the right column.

Place	Latitude and Longitude
____ 1. Algiers, Algeria	a. 34° N, 7° W
____ 2. Cairo, Egypt	b. 33° N, 13° E
____ 3. Rabat, Morocco	c. 30° N, 31° E
____ 4. Tripoli, Libya	d. 37° N, 10° E
____ 5. Tunis, Tunisia	e. 37° N, 3° E

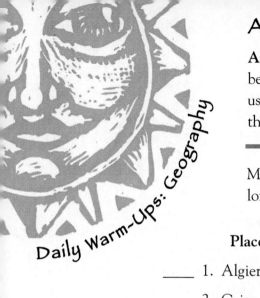

Where Am I?

Use the clues in this paragraph to identify the place being described.

The second-largest country in Africa, this country lies on the shores of the Mediterranean Sea, between Morocco and Tunisia. Its capital is one of the main ports on the North African coast.

This country consists of two distinct regions. In the north, along the Mediterranean, lies a region of wooded hills and fertile plains. The winters along the coast are mild and wet, while the summers are hot and dry. Farmers here grow wheat, citrus fruits, and vegetables.

To the south of this fertile area, a band of mountains crosses the country from west to east. These mountains protect the northern plains from the harsh desert winds of the south. Since very little rain falls in the mountains, life in this area is difficult.

The country south of the mountains is largely desert. In fact, four-fifths of the country is covered by the Sahara. Temperatures here can reach as high as 120° F (49° C). Although the desert area is not suitable for growing crops, it is the source of most of the country's wealth. Oil and natural gas are found under the sands. These make up the country's main exports.

Where am I?

Land Use

Geographers use pie charts to show parts of a whole and how the parts relate to each other. The two pie charts below show land use in two North African countries, Egypt and Morocco. Compare the two charts. Then answer the questions that follow.

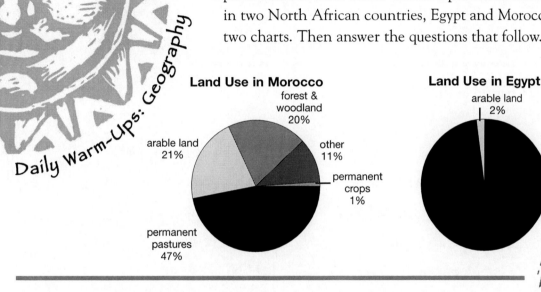

Land Use in Morocco
- forest & woodland 20%
- other 11%
- permanent crops 1%
- permanent pastures 47%
- arable land 21%

Land Use in Egypt
- arable land 2%
- other 98%

1. What are the biggest differences between the two charts?
2. Based on the information you see here, how would you expect the climate and terrain in the two countries to be different?
3. Based on your knowledge of geography, what do you think the land labeled "other" on Egypt's chart is like?

Absolute Location

Absolute location is the exact point on Earth where a place can be found. No two places have the same absolute location. This is usually found by using imaginary lines—latitude and longitude—that mark positions on the surface of the earth.

Match each place named in the left column with its latitude and longitude in the right column.

Place	Latitude and Longitude
___ 1. Bamako, Mali	a. 7° N, 4° W
___ 2. Freetown, Sierra Leone	b. 12° N, 2° W
___ 3. Lagos, Nigeria	c. 13° N, 8° W
___ 4. Monrovia, Liberia	d. 6° N, 11° W
___ 5. Ouagadougou, Burkina Faso	e. 9° N, 13° W

Where Am I?

Use the clues in this paragraph to identify the place being described.

This West African nation is on the Gulf of Guinea, between Benin and Cameroon, with Niger on the north and Chad on the northeast.

The Niger, Africa's third-longest river, enters the country in the northwest. It flows southward through tropical rain forests and swamps to its delta in the Gulf of Guinea. Another major river, the Benue, enters the country from the east. It joins the Niger in the country's center, forming a huge Y. This Y-shape appears on the country's crest.

From ancient times, this country has been a cultural crossroads. Traders from the north and from the interior met here. This has contributed to the country's rich cultural heritage. Today more than 250 ethnic groups live here. At times, this can lead to conflict.

The country is also blessed with rich natural resources. It is one of the world's major oil producers. In the south, rubber and cotton plantations add to the economy. Other industries include mining, especially for tin and coal; textiles; and food products.

Where am I?

Absolute Location

Absolute location is the exact point on Earth where a place can be found. No two places have the same absolute location. This is usually found by using imaginary lines—latitude and longitude—that mark positions on the surface of the earth.

Match each place named in the left column with its latitude and longitude in the right column.

Place	Latitude and Longitude
____ 1. Bangui, Central African Republic	a. 4° N, 12° E
____ 2. Brazzaville, Congo	b. 12° N, 15° E
____ 3. Libreville, Gabon	c. 4° N, 19° E
____ 4. N'Djamena, Chad	d. 4° N, 15° E
____ 5. Yaoundé, Cameroon	e. 1° N, 9° E

Where Am I?

Use the clues in this paragraph to identify the place being described.

This landlocked nation lies south of Libya, north of the Central African Republic, west of Sudan, and east of Niger, Nigeria, and Cameroon. The terrain here is widely varied. In the northwest, the Tibesti Mountains separate it from Libya. South of these mountains, the Sahara Desert covers a broad swath of the country. Then comes an area of broad, dry plains, the Sahel. In the western part of the Sahel lies Lake Chad. This is the most important water body in the Sahel and a significant resource for the country. The lake is rich with fish and provides water for irrigation. South of Lake Chad, the dry plains give way to moist, fertile lowlands. This is the only area suitable for crops. Still, most people in this country depend on agriculture for a living. Cotton is the major export.

The country does have other resources, but these have not yet been developed. One resource that is currently being explored is oil development. The country has extensive reserves of oil around Doba in the south. Development of this oil field should make a difference in the economy here.

Where am I?

Absolute Location

Absolute location is the exact point on Earth where a place can be found. No two places have the same absolute location. This is usually found by using imaginary lines—latitude and longitude—that mark positions on the surface of the earth.

Match each place named in the left column with its latitude and longitude in the right column.

Place	Latitude and Longitude
____ 1. Addis Ababa, Ethiopia	a. 16° N, 33° E
____ 2. Asmara, Eritrea	b. 2° N, 45° E
____ 3. Kampala, Uganda	c. 9° N, 39° E
____ 4. Khartoum, Sudan	d. 0° N, 33° E
____ 5. Mogadishu, Somalia	e. 15° N, 39° E

Where Am I?

Use the clues in this paragraph to identify the place being described.

This country is a sweep of land along the shore of the Red Sea, between Djibouti and Sudan, with Ethiopia to the south. In the northwest, the land is hilly; the southwest is flat or rolling plains. In the east, the land slopes down to a coastal desert plain. This plain includes some of the hottest places on Earth. Summer temperatures average 95–113° F (35–45° C), and rain is rare.

The coastal lowlands are too dry for agriculture. However, this narrow strip is one of the country's greatest economic resources. It gives the country a strategic position on one of the world's busiest shipping lanes—the Red Sea. Access to the Red Sea may also lead to the development of offshore oil and a fishing industry.

This country is Africa's newest republic. Until 1993, it was a province of Ethiopia. Even after it became independent, fighting with Ethiopia continued. Part of the reason for this was geographic. Without this province, Ethiopia is landlocked. It no longer has a port on the Red Sea. Finally, in 2000, the two countries signed a peace agreement. For now, the country is focusing on rebuilding after decades of war.

Where am I?

Absolute Location

Absolute location is the exact point on Earth where a place can be found. No two places have the same absolute location. This is usually found by using imaginary lines—latitude and longitude—that mark positions on the surface of the earth.

Match each place named in the left column with its latitude and longitude in the right column.

Place	Latitude and Longitude
____ 1. Gaborone, Botswana	a. 9° S, 13° E
____ 2. Harare, Zimbabwe	b. 15° S, 28° E
____ 3. Luanda, Angola	c. 18° S, 31° E
____ 4. Lusaka, Zambia	d. 23° S, 17° E
____ 5. Windhoek, Namibia	e. 25° S, 26° E

Where Am I?

Use the clues in this paragraph to identify the place being described.

This island nation lies in the Indian Ocean, about 250 mi (400 km) east of Mozambique. It is the fourth largest island nation in the world.

With a tropical climate along the coast, a temperate one inland, and an arid climate in the south, this country grows a variety of crops. One of its most important crops is cloves, which are the dried flower buds of an evergreen tree. Cloves are used to flavor many different types of foods.

Unfortunately, because most of the people on the island depend on agriculture, farmers have cleared land of trees in order to create more farmland. This has led to severe erosion problems. With no tree roots to hold the soil in place, tropical rains wash it away.

This also affects the habitats of some of the unusual animals found here. Because the island separated from the African mainland about 160 million years ago, animals evolved into some unique species here. Due in part to loss of habitat, more than a hundred of the animal species on the island are considered endangered.

Where am I?

Geography Terms

Different regions have different characteristics. Because of this, each region has its own terminology—words that are used to describe the region.

Each term below describes an aspect of the geography of South Asia. Write a clear definition for each term.

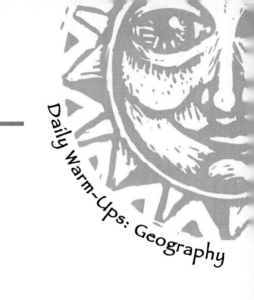

Daily Warm-Ups: Geography

alluvial plain

Deccan

delta

escarpment

ghats

irrigation

monsoon

terai

typhoon

146

© 2003 J. Weston Walch, Publisher

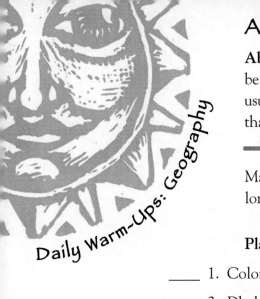

Absolute Location

Absolute location is the exact point on Earth where a place can be found. No two places have the same absolute location. This is usually found by using imaginary lines—latitude and longitude—that mark positions on the surface of the earth.

Match each place named in the left column with its latitude and longitude in the right column.

	Place	Latitude and Longitude
___ 1.	Colombo, Sri Lanka	a. 35° N, 69° E
___ 2.	Dhaka, Bangladesh	b. 19° N, 73° E
___ 3.	Kabul, Afghanistan	c. 28° N, 90° E
___ 4.	Mumbai, India	d. 24° N, 90° E
___ 5.	Thimphu, Bhutan	e. 7° N, 78° E

Relative Location

Every place on Earth has just one absolute location. This is the point on the latitude and longitude grid where it can be found. However, a place can also have a **relative location**. This means where it is relative to other places. If we say a place is north of, south of, or on the shores of, we are giving its relative location. For example, we might describe the relative location of Madras like this: "It is in southeastern India, east of Bangalore, on the Bay of Bengal."

For each place named below, write one or two sentences describing its relative location.

1. Bangalore, India
2. Dhaka, Bangladesh
3. Lahore, Pakistan
4. Mumbai (Bombay), India
5. Qandahar, Afghanistan

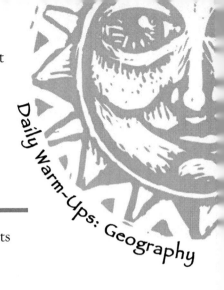

Where Am I?

Use the clues in this paragraph to identify the place being described.

This wedge-shaped country stretches from the Karakoram Mountains in the northeast to the Arabian Sea in the southwest. In the northwest, the Khyber Pass leads through the mountains of the Hindu Kush to Afghanistan. This passage is one of the few ways to cross the mountains that separate central Asia from the Indian subcontinent.

The climate in the north, toward the border with China, is arctic. The climate in much of the rest of the country is hot, dry desert.

Despite the dry climate, the central plain is an important agricultural region. Its name, the Punjab, means "five rivers." The plain is watered by the Indus River and its four major tributaries, the Jhelum, Chenab, Ravi, and Sutlej.

While the Indus makes farming possible, it is also dangerous. After heavy rains in July and August, the river often floods. The country is also subject to frequent earthquakes, especially in the north and west.

Where am I?

What's in a Name?

A **toponym** is a place name. Place names come from a variety of sources. Some places are named after people. Some are named for a distinctive physical feature. Some tell something about the culture of the place.

Each place name listed below means something. Match each name with its meaning. Use your knowledge of geography to help you find the answers.

____ 1. Maldives   a. high land

____ 2. Sri Lanka   b. river

____ 3. Bhutan   c. five rivers

____ 4. Punjab   d. abode of the snow

____ 5. Uttar Pradesh   e. thousand islands

____ 6. Ganges   f. northern province

____ 7. Himalaya   g. beautiful country

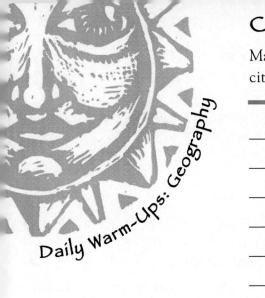

Capital Cities

Match each country name on the left with the name of its capital city on the right.

____ 1. Afghanistan   a. Colombo

____ 2. Bangladesh   b. Dhaka

____ 3. Bhutan   c. Delhi

____ 4. India   d. Islamabad

____ 5. Maldives   e. Kabul

____ 6. Nepal   f. Kathmandu

____ 7. Pakistan   g. Male

____ 8. Sri Lanka   h. Thimphu

Distances

This chart shows distances between some cities in South Asia. To read it, find the starting city in the column to the left of the chart. Then look along the tops of the steps to find the destination city. The box where the row and the column meet gives the distance between those two places. For example, the row for Mumbai and the column for Karachi meet at 547. This means the distance from Mumbai to Karachi is 547 miles.

Study the chart carefully. Then answer the questions that follow.

	Lahore, Pakistan	Calcutta, India	Karachi, Pakistan	Mumbai, India	Columbo, Sri Lanka
Calcutta, India	1082				
Karachi, Pakistan	654	1353			
Mumbai, India	902	1034	547		
Colombo, Sri Lanka	1759	1211	1493	951	
Kathmandu, Nepal	722	410	1145	1000	1478

Distances are shown in miles.

1. How far is it from Karachi to Lahore?
2. Which two cities on this chart are farthest apart? How far apart are they?
3. Which two cities on this chart are closest? How far apart are they?
4. The chart shows distance "as the crow flies," not by road. Do you think distances by road would be about the same as on this chart? Write two or three sentences explaining your answer.

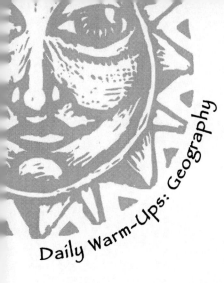

Population of India

The population of India is growing quickly. In 1999, the population reached one billion. Each year, the population grows by about 18 million people.

The line graph below shows India's population from 1950 to 2000. Look carefully at the graph. Then answer the questions that follow.

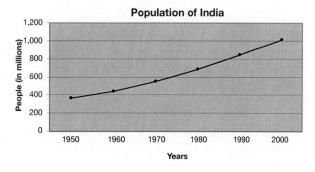

1. What was the approximate population of India in 1950?

2. What was the approximate population of India in 2000?

3. If the population continues to grow at the same rate, about what will the population be in 2050?

4. What kinds of challenges might this large population cause for India? List as many different types of challenges as you can.

Currency

Match each country named below with its currency, listed in the box. Some currencies are used in more than one country.

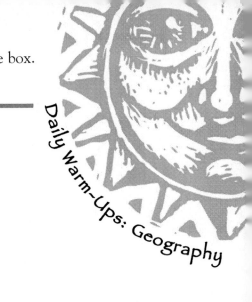

| ngultrum | rufiyaa | rupee | taka |

_____ 1. Bangladesh

_____ 2. Bhutan

_____ 3. India

_____ 4. Nepal

_____ 5. Maldives

_____ 6. Pakistan

_____ 7. Sri Lanka

True or False?

Use your knowledge of the region to decide whether each of these statements is true or false. Write **T** for true or **F** for false on the line beside each statement.

___ 1. The Himalayas, the world's tallest mountains, are getting shorter every year.

___ 2. Ceylon became a republic in 1972 and changed its name to Sri Lanka.

___ 3. Originally, Pakistan included two separate areas, located about 1,000 mi (1,600 km) apart, to the east and west of India.

___ 4. Bangladesh is very sparsely populated.

___ 5. Hinduism is the oldest existing religion on earth.

___ 6. The earliest civilization in South Asia dates to about C.E. 1550.

___ 7. In the fifth century B.C.E., Siddhartha Gautama founded the religion of Buddhism here.

___ 8. Hindi is the only language spoken in India.

___ 9. India is one of the world's top exporters of computer software.

___ 10. People in India are either very rich or very poor; there is no middle class.

Geography Terms

Different regions have different characteristics. Because of this, each region has its own terminology—words that are used to describe the region.

Each term below describes an aspect of the geography of East Asia. Write a clear definition explaining each term.

archipelago

desert

irrigation

loess

monsoon

Pacific Ring of Fire

paddy

tsunami

typhoon

volcano

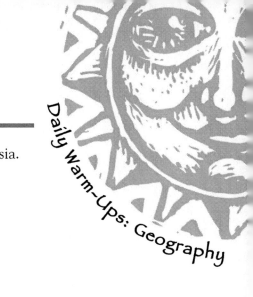

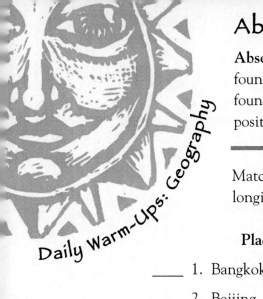

Absolute Location

Absolute location is the exact point on Earth where a place can be found. No two places have the same absolute location. This is usually found by using imaginary lines—latitude and longitude—that mark positions on the surface of the earth.

Match each place named in the left column with its latitude and longitude in the right column.

Place	Latitude and Longitude
____ 1. Bangkok, Thailand	a. 36° N, 140° E
____ 2. Beijing, China	b. 39° N, 126° E
____ 3. Hanoi, Vietnam	c. 40° N, 116° E
____ 4. Pyongyang, North Korea	d. 21° N, 106° E
____ 5. Tokyo, Japan	e. 14° N, 101° E

Relative Location

Every place on Earth has just one absolute location. This is the point on the latitude and longitude grid where it can be found. However, a place can also have a **relative location**. This means where it is relative to other places. If we say a place is north of, south of, or on the shores of, we are giving its relative location. For example, we might describe the relative location of Shanghai like this: "It is in eastern China, southeast of Nanjing, on the Yellow Sea."

For each place named below, write one or two sentences describing its relative location.

1. Hong Kong, China
2. Pusan, South Korea
3. Tokyo, Japan
4. Taipei, Taiwan
5. Vientiane, Laos

Daily Warm-Ups: Geography

What's in a Name?

A **toponym** is a place name. Place names come from a variety of sources. Some places are named after people. Some are named for a distinctive physical feature. Some tell something about the culture of the place.

Each place name listed below means something. Match each name with its meaning. Use your knowledge of geography to help you find the answers.

____ 1. Thailand   a. above the sea
____ 2. Japan   b. northern capital
____ 3. Myanmar   c. land of the free
____ 4. Singapore   d. city of lions
____ 5. Shanghai   e. waterless place
____ 6. Yangon   f. land of the rising sun
____ 7. Hong Kong   g. red hero
____ 8. Ulan Baatar   h. strong and honorable
____ 9. Beijing   i. fragrant harbor
____10. Gobi   j. end of strife

159

© 2003 J. Weston Walch, Publisher

Where Am I?

Use the clues in this paragraph to identify the place being described.

This archipelago has about 14,000 islands; most of them are uninhabited. From west to east, the islands stretch about 3,000 mi (5,000 km) between the Indian Ocean and the Pacific Ocean. The country straddles the equator and lies along major sea lanes.

The islands of this country are rich in natural beauty. But they are also subject to many natural hazards. Many of the islands are mountainous and have active volcanoes. The region is also prone to both earthquakes and tsunamis.

In earlier centuries, parts of this country were known as the Spice Islands. They were the source for spices like pepper, cloves, and nutmeg, which were very valuable in Europe. In fact, these islands were an important reason for the voyages of Christopher Columbus. He was looking for a shorter route to the Spice Islands.

With a tropical monsoon climate in most parts of the country, this is still a rich spice-growing area. However, the country's main exports today are oil, rubber, timber, and coffee.

Where am I?

Capital Cities

Match each country name on the left with the name of its capital city on the right.

____ 1. Cambodia   a. Beijing

____ 2. China   b. Bangkok

____ 3. Indonesia   c. Hanoi

____ 4. Japan   d. Jakarta

____ 5. Laos   e. Kuala Lumpur

____ 6. Malaysia   f. Phnom Penh

____ 7. Mongolia   g. Seoul

____ 8. South Korea h. Tokyo

____ 9. Thailand   i. Ulan Bator

____ 10. Vietnam   j. Vientiane

Unique Nations

Each feature described below is found in one of the countries listed in the box. Match each feature with the country where it is found. Some countries may include more than one feature.

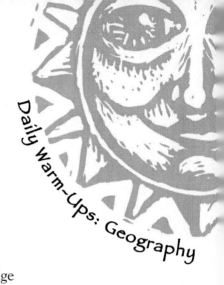

Cambodia	Malaysia
China	Mongolia
Japan	Singapore

_____ 1. World's longest suspension bridge

_____ 2. The only country with only one train station

_____ 3. Country with the world's lowest population density

_____ 4. World's longest strait

_____ 5. World's highest population

_____ 6. Site of the largest known religious building

_____ 7. World's tallest building

© 2003 J. Weston Walch, Publisher

Strait and Narrow

A strait or channel is a narrow body of water that connects two larger bodies of water. There are approximately 200 straits or channels around the world. Some of them are in East Asian waters.

Match each strait in the left column with the bodies of water it connects, listed in the right column.

Strait

____ 1. Strait of Malacca

____ 2. Korea Strait

____ 3. La Pérouse Strait

____ 4. Tatar Strait

____ 5. Taiwan Strait

____ 6. Karimata Strait

____ 7. Makassar Strait

____ 8. Hainan Strait

Bodies of Water

a. Java Sea, South China Sea

b. Java Sea, Celebes Sea

c. South China Sea, East China Sea

d. Andaman Sea, South China Sea

e. Sea of Okhotsk, Sea of Japan

f. Gulf of Tonkin, South China Sea

g. Sea of Japan, East China Sea

Mt. Krakatau

In 1883, Mt. Krakatau, a volcano in Indonesia, erupted. On August 27, the mountain exploded. It spewed out about 5 cu mi (20 cu km) of ash, lava, and debris into the sky in a huge column. Dust from the explosion went as high as 50 mi (80 km) into the atmosphere.

The force of the explosion rocked the seafloor, causing a **tsunami**. The tsunami generated by Mt. Krakatau created a wall of water as high as 140 ft (40 m). When these waves reached the low-lying islands of Indonesia, they swept over everything. Some 165 villages were washed away, and 36,000 people were killed by these waves. The waves were felt as far away as the Arabian Peninsula, more than 4,000 mi (7,000 km) from Mt. Krakatau!

And even then, the effects of the eruption weren't finished. The gas and ash the volcano ejected had an effect on the whole world. The climate was cooled by about .5° C. This may not sound like much. However, in the global climate system, even one small change affects many other things. The drop in temperature caused changes in wind patterns and where storms took place.

How could a volcanic eruption in Indonesia affect the climate of the whole world? Write two or three clear sentences to explain your answer.

Daily Warm-Ups: Geography

The Tibetan Plateau

This area has many nicknames: "Roof of the World," "Land of Snows," "Abode of the Gods." It is a huge plateau. The average elevation is over 13,000 ft (4,000 m)—almost $2\frac{1}{2}$ mi (4 km) above sea level. The climate here is harsh. The year-round average temperature is around 34° F (1° C), with extremes of cold in the winter. Also, the high mountains that surround the plateau block clouds that might carry rain. This makes the land very dry. Only a tiny percentage of the land can be used for growing crops.

Based on what you know about how people interact with their environment, write a clear paragraph describing how people might adapt to living on the Tibetan Plateau.

165

© 2003 J. Weston Walch, Publisher

Bar Chart

Look at the bar chart below. What does it show? How can you tell?

Once you are sure you know what the chart shows, write a descriptive label for the chart. Then write two or three clear sentences explaining what the chart shows.

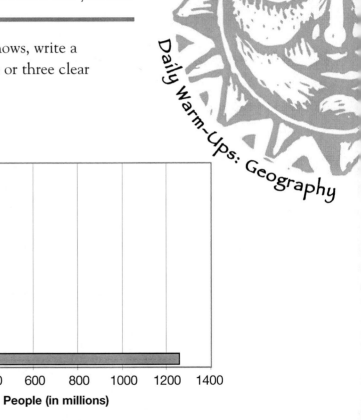

Daily Warm-Ups: Geography

True or False?

Use your knowledge of the region to decide whether each of these statements is true or false. Write **T** for true or **F** for false on the line beside each statement.

___ 1. Thailand means "land of the free."

___ 2. China's Huang He River (Yellow River) has never flooded.

___ 3. Laos was once known as Lan Xang, the Kingdom of a Million Elephants.

___ 4. The full Thai name for the city of Bangkok is Krungthepmahanakhon Amonrattanakosin Mahintharayutthayamahadilok Phopnoppharatratchathaniburirom Udomrathcaniwetmahasathan Amonphimanawatansathit Sakkatthatiyawitsanukamprasit.

___ 5. The Japanese national anthem is the longest in the world.

___ 6. There are no tigers in Indonesia.

___ 7. Although silk was first developed in China, China no longer produces any.

___ 8. With more than 13,600 islands, Indonesia is the world's biggest archipelago.

___ 9. Thailand was a colony of both France and Portugal.

___ 10. Petronas Twin Towers in Kuala Lumpur, Malaysia, is the tallest building in the world.

Currency

Match each country named below with its currency, listed in the box. Some currencies are used in more than one country.

baht	new dong	tugrik
dollar	new kip	won
kina	new riel	yen
kyat	ringgit	yuan
new dollar	rupiah	

_____ 1. Brunei

_____ 2. Cambodia

_____ 3. China

_____ 4. Indonesia

_____ 5. Japan

_____ 6. Laos

_____ 7. Malaysia

_____ 8. Mongolia

_____ 9. Myanmar

_____ 10. North Korea

_____ 11. Papua New Guinea

_____ 12. Singapore

_____ 13. South Korea

_____ 14. Taiwan

_____ 15. Thailand

_____ 16. Vietnam

© 2003 J. Weston Walch, Publisher

Languages

China has a population of more than one billion people, spread out over 3,705,400 sq mi (9,596,960 sq km). Not surprisingly, these people have developed a number of different languages. Language experts say that over 200 languages are spoken in China. One of these, Mandarin Chinese, is the official language.

The table below shows estimates of the number of people who speak 10 languages in China: the 5 languages with the most speakers, and the 5 with the fewest speakers. Your challenge: Choose the best way to show this information in chart or graph form. You don't need to actually create a chart or graph. However, you should write a clear outline describing your approach. It should include: 1. the type of chart or graph you would use; 2. the reason why you think that chart or graph would be best; 3. the scale or scales you would use; and 4. any labels and titles you would use.

Language	Number of Speakers	Language	Number of Speakers
Chinese, Mandarin	867,200,000	Khakas	10
Chinese, Wu	77,175,000	Nanai	40
Chinese, Yue	52,000,000	Manchu	50
Chinese, Jinyu	45,000,000	Samtao	100
Chinese, Xiang	36,015,000	Xiandaohua	100

Geography Terms

Different regions have different characteristics. Because of this, each region has its own terminology—words that are used to describe the region.

Each word below describes an aspect of the geography of the Pacific region. Write a clear definition explaining each word.

archipelago

atoll

coral reef

cyclone

desert

gibber desert

lagoon

mallee

outback

tsunami

volcano

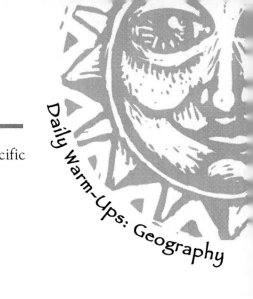

Absolute Location

Absolute location is the exact point on Earth where a place can be found. No two places have the same absolute location. This is usually found by using imaginary lines—latitude and longitude—that mark positions on the surface of the earth.

Match each place named in the left column with its latitude and longitude in the right column.

Place	Latitude and Longitude
____ 1. Canberra, Australia	a. 22° S, 166° E
____ 2. Wellington, New Zealand	b. 8° S, 178° E
____ 3. Nouméa, New Caledonia	c. 1° S, 167° E
____ 4. Nauru	d. 41° S, 175° E
____ 5. Tuvalu	e. 35° S, 149° E

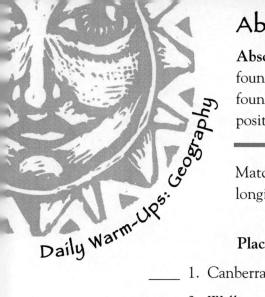

Relative Location

Every place on Earth has just one absolute location. This is the point on the latitude and longitude grid where it can be found. However, a place can also have a **relative location**. This means where it is relative to other places. If we say a place is north of, south of, or on the shores of, we are giving its relative location. For example, to describe the relative location of Hobart, we might say: "It is located in southeastern Australia, on the island of Tasmania, south of Bass Strait."

For each place named below, write one or two sentences describing its relative location.

1. Auckland, New Zealand
2. Melbourne, Australia
3. New Caledonia
4. Perth, Australia
5. Tonga

Where Am I?

Use the clues in this paragraph to identify the place being described.

Although it's January, it is the height of summer here. The temperature is just below freezing—the warmest it ever gets around here. At one point on the continent, the average annual temperature is −71° F (−57° C). The coldest temperature ever recorded here was −128.6° F (−89.3° C).

At 16,860 ft (5,140 m) above sea level, the Vinson Massif is the highest point on the whole continent. With average elevations of 6,500–13,000 ft (2,000–4,000 m), this continent is the highest, on average, in the world. It's also the coldest, windiest, and driest.

There are no real inhabitants here. After all, there is no land suitable for growing crops. However, people come here to learn about the history of our world and the environment. Although minerals such as iron, copper, and gold have been found here, an international agreement prohibits mining any of these minerals.

Where am I?

Capital Cities

Match each country name on the left with the name of its capital city on the right.

____ 1. Australia   a. Apia

____ 2. Cook Islands   b. Avarua

____ 3. Federated States of Micronesia   c. Canberra

____ 4. French Polynesia   d. Honiara

____ 5. Kiribati   e. Manila

____ 6. New Zealand   f. Palikir

____ 7. Northern Mariana Islands   g. Papeete

____ 8. Philippines   h. Saipan

____ 9. Samoa   i. Tarawa

____ 10. Solomon Islands   j. Wellington

Rain in Australia

The chart below shows the annual average rainfall in 12 places in Australia. There are three on Australia's west coast, three on the north coast, three on the east coast, and three inland locations.

Study the chart carefully. What information does it give you? Use this chart and your own background knowledge to guess what kind of population settlement occurs in the different regions. Write two or three clear sentences for your answer.

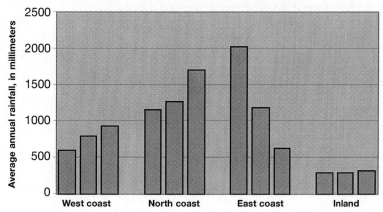

Unique Nations

Each feature described below is found in one of the countries listed in the box. Match each feature with the country where it is found. Some countries may include more than one feature.

Antarctica	Nauru
Australia	New Zealand
Kiribati	

_____ 1. The only country that has every climate type in the world

_____ 2. World's smallest independent republic

_____ 3. Only continent where glaciers are not found

_____ 4. World's largest glacier

_____ 5. World's longest reef

_____ 6. Nation that used to lie on both sides of the International Date Line

Currency

Match each country named below with its currency, listed in the box. Some currencies are used in more than one country.

| dollar |
| franc |
| peso |
| tala |
| vatu |

_____ 1. Australia

_____ 2. Federated States of Micronesia

_____ 3. Fiji

_____ 4. French Polynesia

_____ 5. New Caledonia

_____ 6. New Zealand

_____ 7. Palau

_____ 8. Philippines

_____ 9. Samoa

_____ 10. Solomon Islands

_____ 11. Vanuatu

Thousand Islands

Oceania includes thousands of islands, grouped into a number of different nations and territories. Use your knowledge of the region to identify each island group described below.

---

1. This nation, located at the Eastern Caroline Islands in the northwest Pacific Ocean, has more than 600 islands and 40 volcanoes.

    a. Solomon Islands   b. Micronesia   c. Polynesia   d. Northern Marianas

2. This archipelago of over 7,000 islands, about 500 mi (800 km) southeast of China, is affected by about 15 typhoons each year.

    a. the Philippines   b. Indonesia   c. French Polynesia   d. the Hawaiian islands

3. This group of islands, east of Australia and north of New Zealand, includes more than 300 islands with a total land mass of about 7,000 sq mi (18,000 sq km).

    a. Northern Marianas   b. Marshall Islands   c. Fiji   d. Caroline Islands

4. This group of islands in the Pacific Ocean, halfway between Australia and South America, includes five archipelagos.

    a. French Polynesia   b. Solomon Islands   c. Micronesia   d. Gilbert Islands

True or False

Use your knowledge of the region to decide whether each of these statements is true or false. Write **T** for true or **F** for false on the line beside each statement.

____ 1. Antartica is the only continent without reptiles or snakes.

____ 2. Mt. Uluru/Ayers Rock, in central Australia, looks large because it rises out of a flat plain, but it is actually quite small.

____ 3. More people fit into an average football stadium than have ever been to Antarctica.

____ 4. Australia's Great Barrier Reef is 1,242 mi (2,011 km) long.

____ 5. Most of Australia enjoys a temperate climate, with adequate rainfall for agriculture.

____ 6. Antarctica is the world's largest desert.

____ 7. Because Australia is located so far south, there are no winter sports there, only summer ones.

____ 8. By law, all Australian citizens over the age of 18 must be registered to vote and show up at the poll on election day.

____ 9. Only a tiny fraction of the land in Australia is covered by desert.

____ 10. The Federated States of Micronesia includes many islands, but no volcanoes.

Fill in the Blanks

Choose the best country name to complete each sentence.

Fiji	Nauru	Tuvalu
French Polynesia	New Caledonia	Vanuatu
Kiribati	New Zealand	
Marshall Islands	Tonga	

1. The only country in the world with no official capital is _____.
2. The country of _____ was first settled by the Maoris.
3. The cagou, a small flightless bird, is the emblem of _____.
4. Most of the population of _____ live on the two main islands, Viti Levu and Vanua Levu.
5. Formerly known as the Gilbert Islands, the island nation of _____ won its independence from the United Kingdom in 1979.
6. Bikini Island, in the _____, is still uninhabitable because of U.S. nuclear weapons testing.
7. The only monarchy in Polynesia is the archipelago of _____.
8. With no point on its nine islands rising more than 15 feet (4.5 m) above sea level, the nation of _____ could be swamped if sea levels rise.

Answer Key

1. **Mercator**—<u>Advantages</u>: shapes are accurate and well-defined; all intersections of latitude and longitude maintain angular relationships, making it useful for navigation purposes; directions are represented faithfully; easy to plot a course for navigation using this projection; accurate in the region of the equator. <u>Disadvantages</u>: sizes and areas are very distorted; to create a flat representation of the globe, the Mercator projection spreads the image near the poles, making areas near the poles look much bigger than they really are; distances only true along equator; poles are not shown; not a good projection for global viewing. **Plate Carée**—<u>Advantages</u>: better balance of size and shape in high latitudes than Mercator projection; reduced exaggeration of landmasses; distances true along all parallels and along central meridian; distortion low along equator and in temperate areas, where majority of population lives; scale true along all meridians (it is equidistant); valuable for showing entire world. <u>Disadvantages</u>: Distortion of both shape and area increases with distance from equator. Distortion is greatest near the poles.

2. **alluvial plain**—land along a riverbank where rich sediment from the river is deposited as soil; **aquifer**—underground reservoir of water contained within a porous, water-bearing rock layer; **bay**—an inlet of the sea or other body of water; **delta**—lowland at the mouth of a river, often triangular in shape, formed when the slowing current causes the river deposit sediment; **fjord**—narrow, steep-sided, elongated, and inundated coastal valley deepened by glacier ice that has since melted away, leaving the sea to penetrate the valley; **floodplain**—a flat, low-lying area, near a river or stream, that is subject to flooding; **water table**—the upper level of the area under the ground

Daily Warm-Ups: Geography

where the soil or rock is completely saturated with water
3. **altitude**—vertical distance above sea level; **cartogram**—a "map" that is actually a diagram used to present statistical information, such as a map that shows population by making the size of each country proportional to its population; **contour line**—a line on a map that connects all points that have the same elevation; **equator**—an imaginary line on the earth that has a latitude of 0 degrees; **latitude**—lines of latitude are parallels that are aligned east–west across the globe, from 0° latitude at the equator to 90° north and south latitude at the poles; **longitude**—angular distance (0° to 180°) east or west as measured from the prime meridian (0°) that passes through the Greenwich Observatory in London, England; **map**—a two-dimensional graphical abstraction of the real world, used to show both political and physical features of the earth's surface; **meridian**—line of longitude, aligned north–south across the globe; together with parallels of latitude, they form the global grid system; **parallel**—an east–west line of latitude that is intersected at right angles by meridians of longitude; **projection**—cartographic process used to represent the earth's three-dimensional surface on a two-dimensional map; **scale**—the relationship between the distance shown on a map and the actual distance on the earth's surface; **topographic map**—detailed two-dimensional representation of the natural and human world; usually uses contour lines and a system of symbols to show vertical features on the flat surface of a map
4. 1. b; 2. a; 3. c; 4. c; 5. a; 6. b; 7. c
5. Climate is about long-term records, trends, and averages; weather is the day-to-day

experience. Weather may change on a daily basis—or even more quickly, in some areas; climate changes over a long time. Weather is a description of atmospheric conditions at a certain point in time, while climate is a description of the average conditions—the conditions that can be expected, whether or not those conditions are present at a given time.
6. 1. latitude; 2. longitude; 3. hemisphere; 4. equator; 5. grid
7. Answers will vary; the following is a sample answer. **canyon**—the Grand Canyon, formed by erosion by a river; **delta**—the Mississippi River Delta, formed by deposition of soil carried on a river; **mountain**—Rocky Mountains, formed by upward movement of rock in the earth's crust; **plain**—the Great Plains of the United States, formed by glacial action during the Ice Age; **plateau**—the Colorado Plateau, formed by an uplifting of the earth's crust; **valley**—Monarch Pass, Colorado, formed by glacial action
8. **Push factors**—drought; famine; lack of jobs; overpopulation; war; persecution; human rights violations; political repression; government actions to depopulate or shift ethnic, religious, or other composition of an area; natural disasters (earthquakes, volcanic eruptions, floods, hurricanes); manmade disasters (nuclear accidents, pollution soil erosion, desertification)
Pull factors—economic betterment; educational opportunities; higher standard of living; family reunification; job opportunities; higher wages; better services (schools, housing); better medical facilities
9. I. Aquatic, A. freshwater, 1. ponds and lakes, 2. streams and rivers, 3. wetlands, B. marine, 1. oceans, 2. coral reefs, 3. estuaries

II. Desert, A. hot and dry, B. semiarid, C. coastal, D. cold
III. Forest, A. tropical, B. temperate
IV. Grasslands, A. tropical (savanna), B. temperate, 1. prairie, 2. steppe
V. Tundra, A. arctic, B. alpine

10. 1. avalanche—a sudden downrush of snow or ice; G
2. blizzard—a storm with low temperatures, winds of 32 mph (52 kmh) or higher, and enough snow to reduce visibility to less than 500 ft (150 m); A
3. drought—a freshwater shortage that seriously affects human and animal activity in the region; A
4. earthquake—a shaking of the ground caused by a sudden movement of rock beneath the surface; G
5. flash flood—a sudden flood of great volume, usually caused by heavy rain in the immediate area; A
6. hurricane—a storm above the western Atlantic with rotating winds of more than 74 mph (119 kmh), usually accompanied by heavy rain, thunder, and lightning; A
7. landslide—a downward movement of a mass of soil and rock; G
8. storm surge—water that is pushed toward shore by high wind and changes in air pressure; A
9. thunderstorm—a storm with thunder and lightning, usually also involving strong wind, heavy rain, and sometimes hail; A
10. tornado—a violently spinning column of air, with wind speeds estimated at 250 mph (400 kmh); A
11. tropical storm—a tropical cyclone with winds stronger than 31 mph (50 kmh) but less than 74 mph (119 kmh); A
12. tsunami—a series of large waves, caused

Answer Key

by a sudden displacement of seawater, which produce a destructive surge when they reach land; G
13. volcanic eruption—explosive discharge of lava and gas from a volcanic vent; G

11. The chart shows the distribution of major world religions in 1994. An appropriate label would be "World Religions." Other answers will vary.

12. 1. It shows how many people live, on average, on a square mile of land by continent. 2. No, this is an average figure. More people will live per square mile in cities than in rural areas, and these figures are averaged to find the figures shown on this chart. 3. Asia; 4. Australia and Oceania; 5. Answers will vary. Students should see that a bar chart is an effective way of showing this information.

13. 1. Bolivia; 2. Colombia; 3. Nicaragua; 4. Liechtenstein; 5. Saudi Arabia; 6. Philippines; 7. United States of America

14. 1. F, The average depth of the oceans is about 12,600 ft (3,840 m), with trenches that are more than 18,000 ft (5,500 m) deep. 2. T; 3. F, Coral reefs are found in more than 100 countries. 4. F, At latitude 60° S, you can sail all the way around the world. 5. T; 6. F, The earth's crust is relatively thin, and is composed of separate plates, not a solid whole. As the plates shift, so do the continents. 7. T; 8. T; 9. T

15. 1. Vostok, Antarctica; 2. Plateau Station, Antarctica; 3. Oymyakon, Russia; 4. Verkhoyansk, Russia; 5. Northice, Greenland; 6. Eismitte, Greenland; 7. Snag, Yukon, Canada; 8. Prospect Creek, Alaska, United States; 9. Fort Selkirk, Yukon, Canada; 10. Rogers Pass, Montana, United States

16. 1. Al'Aziziyah, Libya; 2. Death Valley, California, United States; 3. Ghudamis,

Daily Warm-Ups: Geography

Answer Key

Libya; 4. Kebili, Tunisia; 5. Tombouctou, Mali; 6. Araouane, Mali; 7. Tirat Tavi, Israel; 8. Ahwaz, Iran; 9. Agha Jari, Iran; 10. Wadi Halfa, Sudan
17. Answers will vary.
18. 1. Mandarin Chinese; about 900 million; 2. Wu Chinese; about 90 million; 3. They are both Chinese languages.
19. 1. Nile; 2. Angel; 3. Pacific; 4. Baikal; 5. Sahara; 6. Australia; 7. Russia; 8. China
20. **badlands**—an area in South Dakota influenced by wind and water erosion; characterized by deep ravines and gullies, ridges, and a generally barren surface; **cay**—a small, low island; **chinook**—a warm, dry wind along the eastern side of the Rocky Mountains that can result in a rise in temperature of 40° F (20° C) in a quarter of an hour; **coulee**—a dry canyon cut in the western United States; **desert**—an area with little precipitation and sparse vegetation; **fault**—a fracture in the earth's crust accompanied by a displacement of one side of the fracture; **levee**—an embankment built along a river to keep the river from overflowing its banks; **mesa**—a large, flat-topped, steep-sided landform; **piedmont**—an area in the southern states at the base of the Blue Ridge Mountains
21. Answers will vary, but should include at least some of the following: Bible Belt (NC, SC, GA, TN, AR, MO, OK, KS); Corn Belt (IL, IN, IA, NE); Great Lakes (IL, IN, MI, MN, NY, OH, PA, WI); Great Plains (CO, IA, KS, MN, MO, MT, NE, NM, ND, OK, SD, TX, WI, WY); Mid-Atlantic (DE, MD, NY, NJ, PA, DC); Midwest (IL, IN, IA, MI, MN, MO, NB, OH, WI, KS); New England (CT, ME, MA, NH, RI, VT); Northeast (CT, ME, MA, NH, NJ, NY, PA, RI, VT); Northwest (AK, ID, MT, OR, WA, WY); Rocky Mountains

Daily Warm-Ups: Geography

Answer Key

(AZ, CO, ID, MT, NV, NM, UT, WY); Rust Belt (IL, IN, MI, NJ, OH, PA); South (AL, AR, FL, GA, KY, LA, MS, NC, OK, SC, TN, TX, VA, WV); Southeast (AL, AR, FL, GA, KY, LA, MS, MO, NC, SC, TN); Southwest (AZ, CA, NV, NM, TX); Sun Belt (FL, TX, AZ, CA); West (AK, AZ, CA, CO, HI, ID, MT, NV, NM, OR, UT, WY, WA)

22. 1. b; 2. e; 3. d; 4. a; 5. c
23. Sample answers: 1. Boulder: in Colorado, just east of the Rocky Mountains, northwest of Denver, southeast of Laramie, Wyoming, near the Rocky Mountain National Park; 2. Chicago: in Illinois, on the shore of Lake Michigan, south of Milwaukee, Wisconsin, northwest of Indianapolis, Indiana; 3. Miami: in southeastern Florida, on the Atlantic Ocean, northeast of Everglades National Park, south of Boca Raton; 4. New Orleans: in southern Louisiana, on the Mississippi River delta, southeast of Baton Rouge, Louisiana, southwest of Tallahassee, Florida; 5. Seattle: in western Washington state, on Puget Sound, northwest of Mt. Rainier National Park, southeast of Vancouver, Canada
24. Alaska
25. 1. Eastern; 49.6%; 2. "Other" includes Alaskan, Hawaiian, and Atlantic time zones. These areas do not fall into the main time zones because they are geographically distant from the main part of the United States.
26. 1. d; 2. i; 3. f; 4. c; 5. h; 6. a; 7. j; 8. g; 9. b; 10. e
27. 1. d; 2. c; 3. e; 4. b; 5. a
28. 1. e; 2. h; 3. c; 4. j; 5. f; 6. i; 7. a; 8. g; 9. b; 10. d
29. 1. e; 2. b; 3. i; 4. a; 5. d; 6. h; 7. j; 8. f; 9. c; 10. g
30. 1. Minnesota Vikings; 2. San Francisco 49ers; 3. Seattle Mariners; 4. Denver Broncos;

Daily Warm-Ups: Geography

Answer Key

     5. Boston Celtics; 6. Detroit Pistons; 7. Texas Rangers; 7. New York Knicks

31. Georgia, Tennessee, North Carolina, Virginia, West Virginia, Maryland, Pennsylvania, New Jersey, New York, Connecticut, Massachusetts, Vermont, New Hampshire, Maine

32. Texas. The population density in Florida and in Virginia is more than twice as high as in Texas.

33. 1. About 400,000 (The exact figure is 357,232.); 2. About 6,000,000 (The exact figure is 5,984,121.); 3. The greatest proportional growth is between 1900 and 1910, when the population almost doubled (from 518,103 to 1,141,990). 4. The greatest actual growth is between 1990 and 2000, from 4,866,669 to 5,984,121, an increase of 1,117,452.

34. The chart shows the age distribution of the population of Nebraska. It indicates that more than half the inhabitants of the state are between the ages of 18 and 64, and about a quarter are less than 18 years old. A suitable label might be, "Age Distribution of Population of Nebraska, 2000."

35. 1. The graph shows four sets of U.S. trade figures for 1996–2000: exports to Canada, imports from Canada, exports to Mexico, imports from Mexico. 2. The United States imports more from both Canada and Mexico than it exports to them, creating a trade deficit, not a balance. 3. Based on the graph, this goal is being met.

36. 1. a; 2. c; 3. b; 4. d; 5. b

37. 1. F, With 480 inches of rain every year, this is the wettest place in the U.S. The driest place, with 1.5 inches of rain a year, is Death Valley, California. 2. T; 3. F, Cities with state names include Nevada, MO; California, MD; Louisiana, MO; Oregon, WI; Kansas, OK; Wyoming, OH; Michigan, ND; Delaware,

Daily Warm-Ups: Geography

Answer Key

AR; Indiana, PA. 4. F, The northernmost point of the contiguous states is in Minnesota. 5. T; 6. T; 7. F, Q is the only letter that does not appear in any state name. 8. T; 9. T; 10. F, Six European countries are smaller: Andorra, Liechtenstein, Malta, Monaco, San Marino, and Vatican City.

38. 1. Minnesota; 2. Mexico; 3. Missouri; 4. Nevada; 5. Pennsylvania; 6. Kentucky; 7. Connecticut; 8. Colorado

39. **Provinces**: Alberta, British Columbia, Manitoba, New Brunswick, Newfoundland, Nova Scotia, Ontario, Prince Edward Island, Québec, Saskatchewan; **Territories**: Yukon, Nunavut, Northwest Territories

40. 1. c; 2. d; 3. e; 4. a; 5. b

41. Sample answers: 1. Edmonton: in Alberta, western Canada, northeast of Calgary, northwest of Saskatoon; 2. Iqaluit: in Nunavut, on Frobisher Bay, Baffin Island, northwest of Newfoundland, north of Québec; 3. Montréal: in Québec, on the St. Lawrence River, southwest of Québec City, about 25 miles north of the U.S. border; 4. Toronto: in Ontario, on Lake Ontario, southwest of Ottawa and Montréal, northeast of Detroit, Michigan; 5. Winnipeg: in Manitoba, central Canada, southwest of Lake Winnipeg, about 50 miles north of the U.S. border

42. British Columbia

43. 1. h; 2. a; 3. d; 4. f; 5. j; 6. g; 7. e; 8. b; 9. c; 10. i

44. 1. Ontario; 2. Nunavut; 3. Alberta; 4. No; you would need to know the area of each province to determine population density.

45. 1. c; 2. b; 3. e; 4. a; 5. d

46. Answers will vary, but students should recognize that only a country that regularly experiences severe weather would develop such an index.

Daily Warm-Ups: Geography

Answer Key

47. 1. d; 2. h; 3. c; 4. j; 5. a; 6. i; 7. g; 8. e; 9. b; 10. f
48. 1. c; 2. c; 3. a; 4. b; 5. d
49. 1. T; 2. T; 3. F, With almost a million lakes, Canada has more lakes than any other country in the world. 4. T; 5. F, The St. Lawrence Seaway connects the Great Lakes to the Atlantic. 6. T; 7. F, There are six time zones in Canada. 8. F, British Columbia has the most moderate climate of Canada's regions. 9. T; 10. F, Most of Canada's population live within a few hundred miles of the border.
50. 1. Ontario; 2. Newfoundland and Labrador; 3. Yukon; 4. British Columbia; 5. Alberta; 6. Nunavut; 7. Newfoundland and Labrador; 8. Nova Scotia
51. **altiplano**—series of highland valleys and plateaus located in the Andes of Bolivia and Peru; **chaparral**—dense, impenetrable thicket of shrubs or dwarf trees; **cordillero**—related set of separate mountain ranges; **El Niño**—periodic warming of the ocean waters off South America's northwestern coast that affects global weather patterns; **llano**—open grassy plain; **pampas**—vast savanna east of the Andes from Argentina into Uruguay; **pamperos**—violent windstorm in Argentina; **páramos**—series of highland valleys and plateaus in the Andes of Ecuador; **rain forest**—evergreen woodland of the tropics with a continuous leaf canopy and an average rainfall of about 100 in per year; **rain shadow**—the region on the lee side of a mountain or similar barrier where precipitation is less than on the windward side; **sierra**—range of mountains, especially with a serrated or irregular outline
52. 1. b, c, e; 2. a; 3. b, d; 4. b; 5. b, c, d; 6. b, d; 7. a; 8. b, d; 9. a; 10. b; 11. a; 12. b; 13. a; 14. a; 15. a; 16. b, e; 17. b, d; 18. b; 19. b, e; 20. b

Answer Key

53. 1. Bolivia, Peru; 2. Brazil; 3. Brazil; 4. Argentina, Bolivia, Paraguay; 5. Argentina, Uruguay; 6. Ecuador; 7. Ecuador, Peru, Bolivia; 8. Brazil; 9. Mexico
54. 1. c; 2. e; 3. b; 4. a; 5. d
55. Sample answers: 1. Belmopan, Belize: in central Belize, southwest of Belize City, 60 miles from the Caribbean; 2. Buenos Aires, Argentina: in eastern Argentina, where the Uruguay River meets the Atlantic Ocean; 3. Havana, Cuba: in northwestern Cuba, on the Caribbean Sea; 4. Lima, Peru: on Peru's coastal plain, west of the Andes, on the Pacific Ocean; 5. Rio de Janeiro, Brazil: in southeastern Brazil, on the Atlantic Ocean, near the entrance to Guanabara Bay
56. Bolivia
57. 1. d; 2. c; 3. k; 4. b; 5. j; 6. l; 7. h; 8. g; 9. i; 10. a; 11. f; 12. e
58. 1. A little over one million; 2. A little over fifteen million; 3. It increased by the greatest number between 1970 and 1980. 4. The tendency of the population to be concentrated in cities leads to problems with overcrowding, air pollution, water and sanitation facilities, and solid waste disposal.
59. 1. c; 2. c; 3. c; 4. a; 5. d; 6. b; 7. b
60. 1. b; 2. j; 3. h; 4. e; 5. c; 6. g; 7. a; 8. i; 9. f; 10. d
61. Answers will vary but should address the hazards of erosion, as the thin soil is quickly washed away when trees are cleared; damage to river ecosystems as silt deposits impede navigation, increase flooding of low-lying areas, and negatively affect the growth of river plants and fish, which in turn affects wading and fishing birds; increased global carbon dioxide levels, which may contribute to global warming; increased flooding as water is not temporarily held in the forests; and so forth.

Daily Warm-Ups: Geography

62. 1. d; 2. d; 3. a; 4. a; 5. b
63. 1. Mexico; 2. Peru; 3. Chile; 4. Venezuela; 5. Bolivia; 6. Nicaragua; 7. Mexico; 8. Chile; 9. Chile; 10. Mexico
64. 1. F, There are more than 20 volcanoes in Mexico, several of which are currently active. 2. F, Parts of the Amazon have not been explored, and new plants and animals continue to be discovered. 3. T; 4. F, Puerto Rico is east of Cuba, Jamaica, the Bahamas, and Hispaniola. 5. T; 6. F, They are called the Greater and Lesser Antilles. 7. T; 8. F, Bolivia and Paraguay are both landlocked. 9. T; 10. F, The longest mountain range is the Mid-Atlantic Ridge, a submerged range that extends from the Arctic Ocean to the South Atlantic Ocean.
65. 1. Mexico; 2. Panama; 3. Brazil; 4. Argentina; 5. Colombia; 6. Peru; 7. Venezuela; 8. Guyana
66. **bog**—an area of wet, spongy ground; **canton**—a political division or state; one of the 22 Swiss states; **dike**—rock or earthen embankment built to hold back water; **fjord**—narrow, steep-sided, inundated coastal valley deepened by glacier ice that has since melted away, leaving the sea to penetrate the valley; **meseta**—vast plateau in central Spain; **moor**—broad, treeless, rolling land, often poorly drained and having patches of marsh and peat bog; **peninsula**—a strip of land that juts out into an ocean; **polder**—lowland area reclaimed from the sea
67. 1. b, e; 2. a; 3. a, e; 4. a, c; 5. a, c; 6. a, d, e; 7. a; 8. a, d; 9. c; 10. a; 11. a, d, e; 12. a, e; 13. d, e; 14. a; 15. c; 16. a, b, d, e; 17. a, b, d, e; 18. a, c; 19. e; 20. a
68. Answers will vary, but students should recognize that the topography of Western Europe kept peoples separated from each other, which facilitated the development of many different languages and cultures.
69. 1. d; 2. c; 3. e; 4. b; 5. a

Answer Key

70. 1. d; 2. b; 3. a; 4. e; 5. c
71. Sample answers: 1. Hamburg: in northern Germany, south of Kiel, northeast of Bremen, northwest of Berlin, on the Elbe River; 2. Kristianstad: in southern Sweden, southwest of Stockholm, east of Copenhagen, Denmark; 3. Marseille: in southern France, on the Mediterranean, southeast of Montpellier, southwest of Cannes; 4. Naples: in southern Italy, southeast of Rome, north of Sicily, northwest of Mt. Vesuvius; 5. Porto: in northwestern Portugal, north of Fatima, northeast of Lisbon, where the Douro River flows into the Atlantic Ocean.
72. Portugal
73. The climate of Western Europe is influenced by prevailing westerly winds and by the warm waters of the Gulf Stream, a wide ocean current that brings warm water from the Gulf of Mexico north through the Atlantic. Winds blowing over this current are warmed by the water, then carry their warmth inland. These prevailing winds also keep cold arctic air from reaching Western Europe.
74. 1. The southeast coast; 2. The northeast coast; 3. Since the U.K. is at a northerly latitude, days are much shorter in wintertime and longer in summertime than days in more southerly latitudes.
75. 1. a; 2. c; 3. a; 4. d; 5. b
76. 1. j; 2. b; 3. g; 4. a; 5. h; 6. f; 7. c; 8. e; 9. i; 10. d
77. Portugal, Spain, Italy, Greece
78. The economic development of East Germany and West Germany had been very different. West Germany had rebuilt and recovered after World War II, and had become one of the world's strongest economies. East Germany, under Communist control, had a depressed economy and a crumbling

Daily Warm-Ups: Geography

infrastructure. Reunification meant that these two disparate halves needed to be integrated. Government leaders aimed to raise the standard of living in East Germany to that of the West in just a few years, largely by means of state funding. People in the former West Germany found themselves paying higher taxes to finance rebuilding the former East Germany. This caused resentment in the former West Germany. People in the former East Germany were still used to the socialist structure, where the state aims to overcome poverty. Many of them could not adapt to a market economy; the workings of capital were unfamiliar to them. Unemployment rates in the eastern part of Germany are still twice as high as those in the western part, and productivity is much lower than in the west. The standard of living that former East Germans had hoped for still eludes many.

79. Students should note that EU membership includes both economic and political benefits: free trade among member nations; freedom of citizens of member states to work and settle anywhere in the Union; recognition of professional qualifications everywhere within the EU; single currency, which reduces cost of trade among member nations, minimizes currency instability, simplifies projection of future markets, simplifies tourism; improved relationships between member nations; increased choice in consumer goods; and so forth.
80. 1. c; 2. f; 3. h; 4. g; 5. a; 6. d; 7. i; 8. e; 9. b
81. 1. Vatican City; 2. Norway; 3. Iceland; 4. The Netherlands; 5. Finland; 6. Finland; 7. United Kingdom, France; 8. Spain
82. Answers will vary. Students should note that the number and variety of winds suggests an unusual topography, and that the descriptions

… of the winds name both the lake and mountains around it—two of the features described in the introductory text as affecting local winds. Here is a description of Lake Garda's topography: Of glacial origin, it is the largest lake in Italy. It lies in a wide valley, with the Dolomites to the north. The northern part of the lake is long, narrow, and bordered on the west by high, steep banks and on the east by the Monte Baldo ridge. The southern part is wide with low banks and is divided by the narrow Sirmione peninsula. The combination of the mountains, valley, cliffs, and lake surface create unique conditions that lead to the unusual local winds of Lake Garda.

83. 1. It shows that emigration rates have dropped dramatically, while immigration rates have more than doubled. 2. Students should see that the Irish economy has changed dramatically, with greatly improved employment opportunities. 3. Yes, as it enables the viewer to quickly see the general trend of the data.

84. 1. euro; 2. euro; 3. krone; 4. euro; 5. euro; 6. euro; 7. euro; 8. króna; 9. euro; 10. euro; 11. euro; 12. euro; 13. krone; 14. euro; 15. euro; 16. krona; 17. Swiss franc; 18. pound Answers to the question will vary, but students should note that a shared currency facilitates trade among the twelve nations that have adopted the euro.

85. 1. F, Vesuvius is an active volcano. 2. T; 3. F, The sun shines in Greece about 3,000 hours per year, or an average of 8.25 hours every day. 4. T; 5. T; 6. F, No point in Western Europe is more than 300 mi (480 km) from the sea. 7. F, Mt. Blanc, at 15,771 ft (4,807 m), is the tallest in Western Europe. 8. F, Some 40 percent of the land is below sea level. 9. T; 10. T

86. 1. Sweden; 2. Italy; 3. France; 4. Switzerland;

Answer Key

   5. Germany; 6. Ireland; 7. the United Kingdom; 8. Finland
87. **bora**—a cold, violent, northerly wind in the Adriatic area; **chernozem**—fertile black soil; **karst**— landscape composed of limestone features including sinkholes, caves, and underground streams; **loess**—fertile soil made up of small particles transported by the wind to their present location; **puszta**—temperate grasslands of Hungary; **rendzina**—dark grayish-brown, humus-rich soil that develops on limestone
88. 1. b; 2. c; 3. e; 4. a; 5. d
89. Sample answers: 1. Brno: in the southeastern Czech Republic, on the Svratka and Svitava rivers, southeast of Prague; 2. Bucharest: in southeastern Romania, south of the Carpathian Mountains, west of Constanta and the Black Sea; 3. Durres: in western Albania, near the Adriatic Sea, west of Tiranë, southwest of Shkoder; 4. Kraków: in southeastern Poland, north of the Carpathian Mountains, southwest of Warsaw, southeast of Lodz; 5. Maribor: in northeastern Slovenia, northeast of Ljubljana, near the Austrian border.
90. Hungary
91. 1. Albania; 2. Bulgaria; 3. Czech Rep; 4. Macedonia; 5. Romania; 6. Slovenia; 7. Serbia and Montenegro; 8. Slovakia; 9. Poland; 10. Hungary; 11. Croatia; 12. Bosnia/Herzegovina
92. The Danube is used in several ways: as a means of transportation, as a source of hydroelectricity, for drinking water, for irrigation, and for fishing. Its most important use is for freight transportation, as it is the only major European river to flow from west to east, connecting the North Sea and the Black Sea. Canals have been built to connect the Danube to the Main, Oder, and Rhine rivers.

Daily Warm-Ups: Geography

Answer Key

In earlier centuries, when the cities along the Danube were built, water transportation was the fastest means available; even overland travel was much more difficult than travel by water. The Danube is also an important power resource; several countries have built dams and hydroelectric power plants along its length. Both Hungary and Bulgaria rely heavily on the river for irrigation water to support their agricultural economies. Finally, the river and its delta provide a marine habitat for more than 100 species, including some endangered species. Unfortunately, the use of the Danube for transportation and industry has had a negative effect on its viability for drinking water, fishing, and irrigation, as raw sewage, chemicals from agricultural runoff, waste from factories, and oil from ships have polluted the water.

93. 1. b; 2. d; 3. b; 4. a; 5. c
94. 1. There is a wide variation in per capita income. 2. It is about $5,000. 3. It is about $10,000. 4. Per capita income in Slovenia is twice that in the Czech Republic.
95. Student answers will vary, but should reference the fact that the Balkan region has a long history of war and fragmentation. The word was coined in 1919, following World War I, during which the Austro-Hungarian Empire was fragmented and the new nation of Yugoslavia was formed. Since then, the borders in this region have shifted many times.
96. Student answers will vary. However, they should notice that many of the descriptions involve mountains or bodies of water. These orographic features affect the general pressure system to generate local winds.
97. 1. h; 2. e; 3. g; 4. j; 5. d; 6. c; 7. f; 8. i; 9. b; 10. a
98. 1. lek; 2. marka; 3. leva; 4. kuna; 5. koruna;

Daily Warm-Ups: Geography

6. forint; 7. denar; 8. zloty; 9. leu; 10. koruna; 11. tolar; 12. dinar

99. **arctic**—a bitterly cold region near the Arctic Circle; **chernozem**—fertile black soil of the Russian steppes; **permafrost**—a layer of soil that is always frozen; **podzol**—infertile gray soil found in the taiga; **steppe**—mid-latitude grasslands; **taiga**—coniferous forest that begins where the tundra ends and is dominated by spruce, fir, pine, and cedar; also called boreal forest; **tundra**—a treeless plain found in northern latitudes where temperatures are always cool or cold and the only plants are dwarf shrubs, grasses, sedges, lichens, and mosses

100. 1. c, d; 2. c, d; 3. c; 4. a; 5. d; 6. b, c; 7. b, c; 8. a; 9. a; 10. c; 11. a, c; 12. b, c; 13. b, c; 14. c; 15. b, c

101. 1. c; 2. e; 3. d; 4. a; 5. b

102. Sample answers: 1. Almaty: in southeastern Kazakhstan, near the Kyrgyzstan border, southeast of Turkestan, northwest of Mt. Tengri; 2. Gomel: in southeastern Belarus, near the Russian border, southeast of Minsk, northeast of Mozyr; 3. Gyumri: in northwestern Armenia, near the Turkish border, northwest of Yerevan; 4. St. Petersburg: in western Russia, on the Baltic Sea, northwest of Moscow, southwest of Murmansk; 5. Yalta: in southern Ukraine, on the Black Sea, southeast of Kiev and Odessa.

103. 1. d; 2. g; 3. c; 4. a; 5. f; 6. b; 7. e

104. Belarus

105. The increase in car ownership per 1,000 translates into an overall increase of about 247,602 cars—about double the number on Estonia's roads in 1990. However, in 2001, Estonia only had about 18,000 mi (29,200 km) of paved roads; to accommodate the increase in private cars, the infrastructure will need to

Answer Key

be upgraded and expanded. On the mainland, the boggy, wooded nature of the terrain will make this challenging. To build roads on the islands, equipment and materials may need to be ferried across from the mainland. Also, the increase in cars will lead to an increase in CO_2 emissions, adding to the pollution issues Estonia already faces—air pollution from power plants in the northeast that burn shale oil, and soil and groundwater contamination at former Soviet military bases.

106. 1. 10:13 A.M. 2. 6:53 A.M. 3. You would arrive in Vladivostok at 23:53 on June 17 by Moscow time, but it would be 6:53 on the morning of June 18 by Vladivostok time.

107. Topographically, Ukraine is almost entirely gently rolling plains; a wide belt of rich chernozem—fertile humus soil—covers nearly two thirds of the country. Nearly 3,000 rivers flow through Ukraine. Also, the climate is relatively moderate; in July, the hottest month, the daily high averages 73° F (23° C), while January, the coldest month, averages around freezing. This combination of topography, soil, water, and climate makes Ukraine ideal for growing cereal crops such as wheat, barley, rye, and oats.

108. 1. c; 2. a; 3. b

109. Because the graph does not give specific figures, students will not be able to determine the exact dates of each event, but they should be able to identify approximate dates.
1. The law was passed in 1947. Based on the graph, students should guess the mid 1940s as the date.
2. Poaching began to increase in 1991–1992, after the collapse of the Soviet Union; as many as 60 or 70 animals were killed. Between 1991 and 1996, at least 180 tigers were poached.

Daily Warm-Ups: Geography

Answer Key

3. Anti-poaching brigades were set up in 1993 and 1994, and brought poaching under control. Fewer than 20 tigers a year were poached in 1995 and 1996. In August 1995, Russia's Decree 795 declared the Siberian tiger one of the country's most valuable natural objects and specified a federal conservation program.

110. 1. a; 2. d; 3. g; 4. i; 5. f; 6. j; 7. e; 8. b; 9. c; 10. h

111. 1. dram; 2. manat; 3. ruble; 4. kroon; 5. lari; 6. tenge; 7. som; 8. lat; 9. litas; 10. leu; 11. ruble; 12. somoni; 13. manat; 14. hryvnia; 15. sum

112. Answers will vary, but may resemble the following: In the 1700s, all transportation was either by sea or over land, on foot or by horse. There were no trains, cars, long-distance trucks, or airplanes. Russia's harsh climate meant that, once winter set in, the country was essentially cut off from the rest of the world. The snow and cold made land travel difficult, and the ports could not be used in winter. A port on the Baltic Sea gave Russia year-round access to the rest of the world. **Bonus question**: The city was originally named St. Petersburg, to honor the czar who founded it, Peter the Great. During World War I, some Russians felt that "Petersburg" had too German a sound for a Russian city, and the name was changed to "Petrograd." In 1924, after the Russian Revolution, the name was changed to "Leningrad." Then, instead of honoring a czar, it honored a Communist hero. It kept this name for almost 70 years. In 1991, after the collapse of the Soviet Union, a city-wide referendum was held to see what the people wanted their city called. The choice: St. Petersburg. And so the name has gone full circle.

113. 1. T; 2. F, More than 25 percent of the world's forests are in Siberia. 3. F, The climate in this region is damp and cool. 4. F; 5. T; 6. T; 7. T;

Daily Warm-Ups: Geography

Answer Key

8. F, Until 1991, the countries in this region were part of the U.S.S.R. 9. F, Because so much of this region is either arctic or desert, the average population density is low. 10. F, This region combines more than 100 ethnic groups that speak more than 200 different languages.

114. **desert**—an area with little rain, or where evaporation is greater than precipitation; very little grows here. **irrigation**—the artificial watering of croplands; **monsoon**—seasonal winds over the Arabian Sea that blow for six months from the northeast and for six months from the southwest; **oasis**—a fertile place in a desert, usually where groundwater is available near the surface; **wadi**—a depression in a desert region, often the bed of a stream that is dry except in the rainy season.
115. 1. d; 2. e; 3. c; 4. a; 5. b
116. 1. Abu Dhabi: in north central United Arab Emirates, on the Persian Gulf, southwest of Dubai; 2. Beirut: on Lebanon's west coast, on the Mediterranean Sea, north of Sidon and Tyre; 3. Kuwait: in eastern Kuwait, on the Persian Gulf; 4. Manama: in northern Bahrain, on the Persian Gulf; 5. Tel Aviv: in western Israel, south of Haifa, on the Mediterranean Sea
117. Turkey
118. 1. j; 2. c; 3. f; 4. b; 5. d; 6. g; 7. h; 8. e; 9. a; 10. i
119. Student answers will vary, but should note that Istanbul's location places it at the crossroads of both land and sea traffic between Europe and Asia. In earlier periods, this strategic location meant that the ruler who held Istanbul could stop people from moving between the two regions, whether they traveled for economic reasons, military reasons, or political reasons. It also meant that the rulers

Answer Key

of Istanbul could levy taxes and tolls on merchants trying to transport goods through the region. Even today, the narrow span of the Bosporus means that Istanbul could easily stop all maritime traffic between the Mediterranean and the Black Sea.

120. 1. e; 2. c; 3. a; 4. f; 5. b; 6. d
121. Saudi Arabia, Yemen, Oman, the United Arab Emirates, Qatar, Bahrain, and Kuwait
122. 1. The general trend is an increase in population. 2. The greatest percentage of increase is in Oman; the rate of increase is 3.5%. 3. The greatest numerical increase will be in Turkey, because it starts with the greatest number of people and has a higher rate of population increase than Iran, the only other country with a population close to Turkey's.
123. 1. afghani; 2. dinar; 3. pound; 4. rial; 5. dinar; 6. new shekel; 7. dinar; 8. dinar; 9. pound; 10. dinar; 11. rial; 12. riyal; 13. riyal; 14. pound; 15. lira; 16. dirham; 17. rial
124. The canal creates a shortcut for ships going from Europe to Asia. Instead of having to sail all the way around Africa, they can just sail the 100 mi (163 km) of the canal.
125. 1. a; 2. d; 3. c; 4. a; 5. c
126. **cataract**—a large waterfall; **delta**—fan-shaped area at the mouth of a river, formed by material that has been carried downstream and dropped in quantities larger than can be carried off by tides or currents; **desert**—area with little precipitation and sparse vegetation; **equator**—zero degrees latitude, divides the earth into the Northern and Southern hemispheres; **harmattan**—a dry, dust-bearing desert breeze blowing from the northeast or east over west Africa, usually between late November and mid-March; **sahel**—dryland transition zone between the Sahara Desert to the north and tropical forests to the south; **savanna**—a plain covered with coarse grasses and a few scattered trees; **sirocco**—a warm,

Answer Key

south or southeast wind that occurs ahead of an eastward-advancing depression across North Africa; **veldt**—extensive grassland region of eastern and southern Africa, characterized by a flat terrain and a mixture of trees and shrubs.

127. Students should name at least three of the following for each region.
North Africa—Algeria, Egypt, Libya, Morocco, Tunisia; **West Africa**—Benin, Burkina Faso, Cape Verde, Côte d'Ivoire, Gambia, Ghana, Guinea, Guinea Bissau, Liberia, Mali, Mauritania, Niger, Nigeria, Sénégal, Sierra Leone, Togo; **East Africa**—Comoros, Djibouti, Eritrea, Ethiopia, Kenya, Madagascar, Mauritius, Seychelles, Somalia, Sudan, Tanzania, Uganda; **Central Africa**—Burundi, Cameroon, Central African Republic, Chad, Democratic Republic of the Congo, Equatorial Guinea, Gabon, Republic of the Congo, Rwanda, São Tomé and Principé; **Southern Africa**—Angola, Botswana, Lesotho, Malawi, Mozambique, Namibia, South Africa, Swaziland, Zambia, Zimbabwe

128. The graph shows the 10 most populated countries in Africa, in order from highest to lowest. An appropriate label might be, "Ten Most Populated Nations in Africa."

129. 1. Namibia; 2. Egypt; 3. Liberia; 4. Libya; 5. Sudan; 6. South Africa; 7. Egypt; 8. Tanzania; 9. Egypt

130. 1. f; 2. c; 3. h; 4. a; 5. j; 6. g; 7. d; 8. b; 9. i; 10. e

131. Answers will vary. Possible solutions include: help drought-prone nations develop irrigation works, water supply systems, and dry-land farming methods; resettle displaced persons on newly reclaimed land and help provide employment opportunities; support rural

Daily Warm-Ups: Geography

Answer Key

development through food-for-work and food-for-training programs.

132. One reason is animals' susceptibility to temperature extremes. Most animals can only function within a narrow temperature range. When temperatures are too hot or too cold, the animal cannot survive. Daytime temperatures in the desert are often too high for most animal species. However, because the desert has no vegetation or groundwater to retain heat, the desert cools off quickly once the sun sets. Thus, animals that are only active after sunset are exposed to less extreme heat. Also, desert animals have limited access to water. During the day, water is lost quickly by evaporation. At night, the water loss is less.

133. Sample answers: 1. Abuja: in central Nigeria, northeast of Lagos, southwest of Kaduna; 2. Antananarivo: in central Madagascar, southwest of Toamasina, northeast of Fianarantsoa; 3. Bamako: in southwestern Mali, southeast of Lake Bamako, on the Niger River; 4. Brazzaville: in southern Congo, on the Congo River, northeast of Pointe-Noire; 5. Dodoma: in east-central Tanzania, east of Lake Tanganyika, southeast of Lake Victoria, northwest of Dar es Salaam; 6. Khartoum: in central Sudan, on the Nile River, southwest of Port Sudan; 7. Kigali: in central Rwanda, east of Lake Kiva, northeast of Butare, west of Kagera National Park; 8. Monrovia: in western Liberia, on the Atlantic Ocean, at the mouth of the St. Paul River; 9. Niamey: in southwestern Niger, on the Niger River, near the border with Burkina Faso; 10. Tunis, Tunisia: in northern Tunisia, on the Mediterranean Sea, northwest of Kairouan, south of Carthage

134. 1. b; 2. h; 3. d; 4. c; 5. f; 6. a; 7. j; 8. g; 9. i; 10. e
135. 1. e; 2. c; 3. a; 4. b; 5. d

Daily Warm-Ups: Geography

Answer Key

136. Algeria
137. 1. Egypt has no permanent crops, permanent pastures, or forests and woodlands, but almost 70% of the land in Morocco falls under those headings. Also, only 2% of Egypt's land is arable as compared to 22% of Morocco's.
2. Answers will vary, but students should see that the climate in Egypt is likely to be much drier than that in Morocco; otherwise, Egypt would probably have some forested land or pasture. 3. The land labeled "other" is desert, as most of Egypt is covered by the Sahara.
138. 1. c; 2. e; 3. a; 4. d; 5. b
139. Nigeria
140. 1. c; 2. d; 3. e; 4. b; 5. a
141. Chad
142. 1. c; 2. e; 3. d; 4. a; 5. b
143. Eritrea
144. 1. e; 2. c; 3. a; 4. b; 5. d
145. Madagascar
146. **alluvial plain**—land surface produced by the deposition of clay, silt, sand, gravel, and other loose material, by streams or other bodies of running water; **Deccan**—great triangular plateau of India, much of which is a fertile rice-growing region; **delta**—fan-shaped area at the mouth or lower end of a river, formed by eroded material that has been carried downstream and dropped in quantities larger than can be carried off by tides or currents; **escarpment**—steep slope or cliff that separates two relatively level areas of distinctly different elevations; **ghats**—ranges of hills that run parallel to India's eastern and western coasts and separate the fertile coastal strips from the interior; **irrigation**—the artificial watering of croplands; **monsoon**—wind system in Southeast Asia that changes direction seasonally, creating wet and dry seasons; **terai**—belt of swampy grass jungle between the

Daily Warm-Ups: Geography

Answer Key

Himalayan foothills and the plains of India; **typhoon**—Pacific tropical cyclone with winds of at least 74 mph (119 km/hr); the term "hurricane" is used for Atlantic cyclones.

147. 1. e; 2. d; 3. a; 4. b; 5. c
148. Sample answers: 1. Bangalore: in Karnataka Province, southern India, east of Madras, south of Hyderabad; 2. Dhaka: in central Bangladesh, on the Jamuna River, northwest of Narayangan; 3. Lahore: in northeastern Pakistan, southeast of Islamabad, near the Indian border; 4. Mumbai (Bombay): in western India, northwest of Bangalore, southwest of Calcutta, on the Arabian Sea; 5. Qandahar: in southern Afghanistan, southwest of Kabul.
149. Pakistan
150. 1. e; 2. g; 3. a; 4. c; 5. f; 6. b; 7. d
151. 1. e; 2. b; 3. h; 4. c; 5. g; 6. f; 7. d; 8. a
152. 1. 654 miles; 2. Colombo and Lahore; they are 1,759 miles apart; 3. Kathmandu and Calcutta; they are 410 miles apart; 4. No, distances by road will be longer—often much longer, as roads need to go around physical features like lakes and mountains.
153. 1. About 400 million (actual figure: 369,880,000); 2. About one billion (actual figure: 1,017,645,000). 3. Answers will vary. Analysts project that the population of India in 2050 will be about 1.7 billion. 4. Challenges include: demand depleting natural resource base; deforestation; erosion; falling water tables; educating 18 million more children each year; providing jobs for the 10 million new entrants into the job market each year.
154. 1. taka; 2. ngultrum; 3. rupee; 4. rupee; 5. rufiyaa; 6. rupee; 7. rupee
155. 1. F, Because of the pressure exerted by two continental plates pushing against each other, the Himalayas are in fact growing by about half an inch every year. 2. T; 3. T; 4. F, With more than 1,970 people per square mile, Bangladesh is the most densely populated

Answer Key

non-island region in the world. 5. T; 6. F, The Indus River Valley is the site of one of the world's earliest civilizations; the earliest cities in this region were built about 5,000 years ago. 7. T; 8. F, While Hindi is India's recognized official language, the Indian Constitution recognizes many different national languages; in all, some 1,600 different dialects are spoken in India. Besides Hindi and English, the other recognized languages are Assamese, Bengali, Gujarati, Kannada, Kashmiri, Konkani, Malayalam, Marathi, Oriya, Punjabi, Sanskrit, Sindhi, Tamil, Telugu, and Urdu. 9. T; 10. F, India has the largest middle class in the world, estimated at 300 million.

156. **archipelago**—a chain or set of islands grouped together; **desert**—an area with little rain, or where evaporation is greater than precipitation; very little grows here. **irrigation**—the artificial watering of croplands; **loess**—fertile soil made up of small particles transported by the wind to their present location; **monsoon**—a wind system that changes direction seasonally, creating wet and dry seasons; **Pacific Ring of Fire**—zone of crustal instability along tectonic plate boundaries, marked by earthquakes and volcanic activity, that rings the Pacific Ocean basin; **paddy**—a small plot of land that can be flooded for growing rice; **tsunami**—a wave or series of waves in a body of water caused by a sudden disturbance that displaces water, such as an underwater earthquake; **typhoon**—tropical storm of the Pacific Ocean that has winds of at least 74 mph (119 km/h); **volcano**—naturally occurring opening in the Earth's surface through which molten, solid, and gaseous materials erupt. Volcanic eruptions inject large quantities of dust, gas, and aerosols into the atmosphere.

157. 1. e; 2. c; 3. d; 4. b; 5. a

Answer Key

158. Sample answers: 1. Hong Kong: in southern China, northeast of Macao, on the South China Sea; 2. Pusan: in southeastern South Korea, southeast of Seoul, on the Sea of Japan; 3. Tokyo: in eastern Japan, on Honshu island, on Tokyo Bay, northeast of Yokohama; 4. Taipei: in northern Taiwan, west of Chilung, southwest of Mt. Tatun and Mt. Chihsing; 5. Vientiane: in northwestern Laos, on the Mekong River, near the Thailand border, southwest of Hanoi, northeast of Bangkok.
159. 1. c; 2. f; 3. h; 4. d; 5. a; 6. j; 7. i; 8. g; 9. b; 10. e
160. Indonesia
161. 1. f; 2. a; 3. d; 4. h; 5. j; 6. e; 7. i; 8. g; 9. b; 10. c
162. 1. Japan; 2. Singapore; 3. Mongolia; 4. Malaysia; 5. China; 6. Cambodia; 7. Malaysia
163. 1. d; 2. g; 3. e; 4. e; 5. c; 6. a; 7. b; 8. f
164. The explosion sent a huge dust cloud high into the lower stratosphere; this cloud quickly circled the globe. The volcanic particles and aerosols in the haze reflected solar radiation back into space, reducing solar warming of the earth's surface.
165. Answers will vary. One of the greatest challenges to life on the Tibetan Plateau is the elevation. People from lower elevations may develop hypoxia—altitude sickness—at altitudes of about 6,500 ft (2,000 m). Symptoms include an inability to do normal physical activities, such as climbing a short flight of stairs without fatigue; lack of appetite; and difficulty memorizing and thinking clearly. The people who live on the Tibetan Plateau tend to have greater lung expansion capability and to produce more oxygen-carrying red blood cells. Other adaptations to life on the

Answer Key

plateau include developing crops that can grow despite an arid climate and brief growing season; these include barley, wheat, maize, mustard, millet, sorghum, buckwheat, and rice. The staple diet of the Tibetan people is tsampa, a roasted barley dish. Animal protein comes from the yak, a hardy animal raised for its wool, milk, and meat.

166. The chart shows the population, in millions, of 12 East Asian countries in 2000. A suitable label might read, "Population of East Asian Countries."

167. 1. T; 2. F, The Huang He has flooded more than 1,500 times in the last 1,800 years. 3. T; 4. T; 5. F, The Japanese anthem is only four lines long. 6. F; 7. F, China produces about 70 percent of the world's silk supply. 8. T; 9. F, Thailand was never a colony of a European power. 10. T

168. 1. ringgit; 2. new riel; 3. yuan; 4. rupiah; 5. yen; 6. new kip; 7. ringgit; 8. tugrik; 9. kyat; 10. yen; 11. kina; 12. dollar; 13. yen; 14. new dollar; 15. baht; 16. new dong

169. Answers will vary. However, neither a line graph nor a pie chart would be an appropriate way to show this information. One possible solution would be to create two bar charts, one for the top 5 languages, one for the bottom 5; this would allow for the use of two separate scales, as no one scale can express both 10 and 867,200,000 in any meaningful way.

170. **archipelago**—a chain or set of islands grouped together; **atoll**—a circular coral reef that encloses a shallow lagoon; a coral island consisting of a ring of coral surrounding a central lagoon; atolls are common in the Indian and Pacific oceans. **coral reef**—marine ridge, found in coastal zones of warm oceans, produced by the exterior skeletons of colonies of tiny animals; **cyclone**—tropical storm of the

Daily Warm-Ups: Geography

Pacific Ocean that has winds of at least 74 mph (119 kmh); known as hurricane in the western Atlantic; **desert**—an area with little rain, or where evaporation is greater than precipitation; very little grows here. **gibber desert**—large areas in Australia covered by small, rounded pebbles, rocky ranges, and low, scrubby vegetation; **lagoon**—small, shallow body of water between a barrier island or a coral reef and the mainland, also a small body of water surrounded by an atoll; **mallee**—grassy, open woodland habitat characteristic of many semi-arid parts of Australia; **outback**—arid, sparsely settled central and western plains and plateaus of Australia; **tsunami**—a wave or series of waves in a body of water caused by a sudden disturbance that displaces water, such as an underwater earthquake; **volcano**—naturally occurring opening in the earth's surface through which molten, solid, and gaseous materials erupt; volcanic eruptions inject large quantities of dust, gas, and aerosols into the atmosphere.

171. 1. e; 2. d; 3. a; 4. c; 5. b
172. Sample answers: 1. Auckland: toward the north of North Island, north of Wellington, northeast of Christchurch; 2. Melbourne: in southeastern Australia, southwest of Canberra, on the Yarra River; 3. New Caledonia: island in the South Pacific, northeast of Australia, northwest of New Zealand, southwest of Fiji; 4. Perth: in southwestern Australia, where the Swan River flows into the Indian Ocean, west of the Darling Range; 5. Tonga: island in the South Pacific, northeast of New Zealand, southwest of American Samoa, west of French Polynesia, just west of the International Date Line
173. Antarctica
174. 1. c; 2. b; 3. f; 4. g; 5. i; 6. j; 7. h; 8. e; 9. a; 10. d

Answer Key

175. Most precipitation in Australia falls on a narrow coastal strip; the interior of the country is very dry. This is one reason that most major settlements in Australia are on the coasts, not in the interior.
176. 1. New Zealand; 2. Nauru; 3. Australia; 4. Antarctica; 5. Australia; 6. Kiribati
177. 1. dollar; 2. dollar; 3. dollar; 4. franc; 5. franc; 6. dollar; 7. dollar; 8. peso; 9. tala; 10. dollar; 11. vatu
178. 1. b; 2. a; 3. c; 4. a
179. 1. T; 2. F, Ayers Rock/Uluru rises 1000 ft (348 m) above the desert floor and measures 5.5 mi (9.4 km) around. It has been identified as the largest rock in the world. 3. T; 4. T; 5. F, The interior of Australia is hot and dry; about one fifth of the country is desert. 6. T; 7. F, Australia has a number of popular ski resorts. 8. T; 9. F, About one fifth of Australia is covered by its eleven deserts. 10. F, There are about 40 volcanoes in Micronesia's 600 islands.
180. 1. Nauru; 2. New Zealand; 3. New Caledonia; 4. Fiji; 5. Kiribati; 6. Marshall Islands; 7. Tonga; 8. Tuvalu

Turn downtime into learning time!

Other books in the

Daily Warm-Ups series:

- Algebra
- Analogies
- Biology
- Critical Thinking
- Earth Science
- Geometry
- Journal Writing
- Poetry
- Pre-Algebra
- Shakespeare
- Spelling & Grammar
- Test-Prep Words
- U.S. History
- Vocabulary
- Writing
- World History